Explanation, Laws, and Causation

Scientific explanation, laws of nature, and causation are crucial and frontier issues in the philosophy of science. This book studies the complex relationship between the three concepts, aiming to achieve a holistic synthesis about explanation–laws–causation.

By reviewing Hempel's scientific explanation models and Salmon's three conceptions – the epistemic, modal, and ontic conception – the book suggests that laws are essential to explanation and that our understanding of laws will help solve the problems of the latter. Concerning the nature of laws, this book tackles both the problems of regularity approach and necessitarian approach. It also proposes that the ontological order of explanation should be from events (or processes) to causation, then to regularity (laws), and finally to science system, but the epistemological order should be from science system to laws to explanation and causation. In addition, this book examines the legitimacy of *ceteris paribus* laws, the connection between explanation and reduction, the relation between explanation and interpretation, and some other issues closely related to explanation–laws–causation. This book will attract scholars and students of philosophy of science, natural sciences, social sciences, etc.

Wei Wang is a professor at the Institute of Science, Technology and Society, Tsinghua University. His research interests include general philosophy of science, philosophy of social science, and philosophy of biology.

China Perspectives series

The *China Perspectives* series focuses on translating and publishing works by leading Chinese scholars, writing about both global topics and China-related themes. It covers Humanities & Social Sciences, Education, Media and Psychology, as well as many interdisciplinary themes.

This is the first time any of these books have been published in English for international readers. The series aims to put forward a Chinese perspective, give insights into cutting-edge academic thinking in China, and inspire researchers globally.

For more information, please visit www.routledge.com/series/CPH.

Forthcoming titles:

Return to Scientific Practice – A New Reflection of Philosophy of Science
Tong Wu

Central Issues of Philosophy of Science
Wei Wang

Secret Subversion I: Mou Zongsan, Kant, and Original Confuciaism
Wenming Tang

Secret Subversion II: Mou Zongsan, Kant, and Original Confuciaism
Wenming Tang

Confucius and Modern World
Lai Chen

Explanation, Laws, and Causation

Wei Wang

Routledge
Taylor & Francis Group

LONDON AND NEW YORK

This book is published with financial support from the Chinese Fund for the Humanities and Social Sciences

First published 2017
by Routledge
2 Park Square, Milton Park, Abingdon, Oxon OX14 4RN

and by Routledge
711 Third Avenue, New York, NY 10017

Routledge is an imprint of the Taylor & Francis Group, an informa business

British Library Cataloguing-in-Publication Data
A catalogue record for this book is available from the British Library

Library of Congress Cataloging-in-Publication Data
A catalog record for this book has been requested

ISBN: 978-1-138-84583-1 (hbk)
ISBN: 978-1-315-72790-5 (ebk)

Typeset in Times New Roman
by Apex CoVantage, LLC

Contents

Preface for the English version

Many thanks to Routledge and Taylor & Francis Group for publishing the English version. The Chinese version of this book was finished in draft form in 2010 and was then published in 2011, so I had not tackled currently the most influential account of scientific explanation, James Woodward's manipulationist account. Woodward's *Making Things Happen: A Theory of Causal Explanation* was published in 2003 and received the Lakatos Award in 2005. In this work, he argues against the unificationist account and constructs a comprehensive theory of causal explanation. In response, I am currently writing a paper, coauthored with Professor Chuang Liu at the University of Florida, criticizing Woodard's manipulationist account and calling for peaceful coexistence between Woodard's causal explanation and Philip Kitcher's unificationist model.

This is my first time publishing a book in English with an international publisher. Stacey Carter at Routledge, Taylor & Francis Group, Autumn Spalding at Apex, and Qiuwan Zhuang, Qiaozheng Wang, and Christine Lv at Tsinghua University Press have provided me with great assistance in making this publication a reality.

The translation of this book was greatly supported by my graduates: Jiuheng He for Chapter 1, Jia Liu for Chapter 2, Yichen Luo for Chapter 3, Zhenyu Sun for Chapter 4, Guanyao Hu for Chapter 5, Jiahui Zhu for Chapter 7, and Dian Zeng for Chapter 8. My former graduate Mingjun Zhang, now a PhD student at the University of Pennsylvania, provided most of the English quotations. Ryan Pino, a graduate student in the School of Philosophy at Fudan University, polished the English of the whole book. He is not only excellent at English polishing, but is also experienced at editing, which helped make the text much more fluent. However, I take full responsibility for any mistakes or errors in the book.

The translation and publication of the English version have been sponsored by the Chinese Fund for the Humanities and Social Sciences (No. 16WXZ011).

Preface for the Chinese version

My research on explanation, laws, and causation started from the concept of scientific explanation. Carl Gustav Hempel's explanation models are praised as a paradigm of the philosophical analysis of scientific concepts. However, even his well-articulated explanation models have been met with many counter-examples and problems. I was very curious about this and began to study the issue. In doing so, I came to realize that Hempel's explanation models were also called covering law models, yet he did not provide a satisfactory answer to the nature of laws. Indeed, because the debate surrounding laws of nature remains a controversial topic in the philosophy of science, I wish to combine scientific explanation and laws of nature and then pursue their common answer.

There has been an academic succession that partly explains my interest in studying the explanation–law–causation issue. When I was a visiting fellow at the Center for Philosophy of Science, the University of Pittsburgh, from 2005 to 2006, I met the former director Professor Gerald Massey. Learning that he had taught at Michigan State University in the 1960s and 1970s, I recalled that my PhD supervisor at the Chinese University of Hong Kong, Professor Hsiu-hwang Ho, had pursued his PhD degree at MSU during that time. Sure enough, when I asked Massey whether or not he knew a Taiwanese student, he told me Ho's name precisely; to my surprise, Ho was Massey's PhD student! Then, Massey proudly added one sentence: that he was C. G. Hempel's PhD student. So, academically, I can be considered a great-grandson of Hempel!

In 2004, I received a grant for a Youth Project titled "Scientific Explanation and Laws of Nature," which was supported by the National Social Science Foundation. Then sponsored by the Harvard-Yenching Visiting Scholarship, I visited Harvard University from 2004 to 2005, followed by a visit to the Pitt Center from 2005 to 2006 sponsored by the Center for Philosophy of Science at the University of Pittsburgh. During this visiting period, I focused on the issue of explanation and law. In particular, the

University of Pittsburgh, which has been ranked No. 1 in the general philosophy of science by the Philosophical Gourmet Report, was instrumental in my research, as many great philosophers dealing with this issue have taught or still teach there, including C. G. Hempel, the creator of explanation models, Wesley Salmon, the main proposer of causal mechanism model, and John Earman, a top philosopher of physics and the most important opponent of *ceteris paribus* laws. I made good use of the opportunity to collect much of the literature on the topics and asked several top philosophers for advice, and while at the Pitt Center, I finished the drafts of several of the chapters found within this manuscript, such as "The Conceptions of Scientific Explanation and Approaches to Laws of Nature" and "Is There Such a Thing as a *Ceteris Paribus* Law?"

After returning to China in July of 2006, on behalf of Tsinghua University, I began to organize the 13th International Congress of Logic, Methodology, and Philosophy of Science, which is hosted every 4 years and is the top conference of logic and philosophy of science in the world. Even though it was a huge conference, the budget was limited, and there were only a few student helpers. As the general secretary of the Local Organizing Committee and a member of the General Program Committee, I tried my best and worked very hard to organize the Congress, which was finally hosted successfully in Beijing from August 9 to August 15, 2007.

However, due to the great amount of time and effort I had to put into organizing the 13th Congress, I had to interrupt this research project. Still, through the Congress, I became acquainted with many distinguished international scholars. Among these scholars were Professor Clark Glymour at Carnegie Mellon University and Professor Dag Westerståhl at the University of Gothenburg, with whom I later coedited the *Logic, Methodology and Philosophy of Science – Proceedings of the Thirteenth International Congress*, which was published by King's Publications in 2009. The volume collected the newest papers from top logicians and philosophers of science, and as the correspondence editor, I became more familiar with the frontiers in the philosophy of science.

I also established long-term cooperation between the Center for Philosophy of Science at the University of Pittsburgh and the Institute of Science, Technology, and Society at Tsinghua University, and founded the Pitt-Tsinghua Summer School for Philosophy of Science. Since 2008, we have invited Professor Sandra Mitchell to teach "Philosophy of Biology," Professor Merrilee Salmon to teach "Philosophy of Social Science," and Professor Clark Glymour to teach "Computation and Causation." These summer schools have not only promoted academic exchange in the philosophy of science, but they have also benefitted my research greatly.

When I first started this research project, I mostly paid attention to the relationship between scientific explanation and laws of nature; however, as later research has shown me, the concept of causation is also important for understanding this relationship between explanation and laws. Recognizing this, I added causation to the equation and finally tried to solve the complex, three-body problem cluster of explanation–law–causation. In recent years, I have finished research on some specific topics within this cluster, with seven *CSSCI* journal papers published, two papers forthcoming, and the draft of *Explanation, Laws, and Causation* finished in 2010.

This book tackles several topics, and the author tries to provide his own synthesis. Concerning explanation models and their problems, the author reviews six decades of work on scientific explanation, not only clarifying the new conceptions and models of scientific explanation, but also pointing out their problems. As I think laws of nature are essential to scientific explanation, so I argue that our understanding of laws will be helpful for solving the problems of scientific explanation.

Concerning the nature of laws, I tackle both the problems of the regularity approach and logical difficulties of the necessitarian approach. I discuss how, while Bas van Fraassen and Ronald Giere refuse to utilize laws in science, they merely change the problem of laws into others such as symmetry or equations, rather than solving or dissolving the problem. In contrast, Sandra Mitchell's account of laws is more plausible, but her elimination of the dichotomy of scientific laws and accidental generalizations is not convincing. I suggest that, ontologically, the necessity of laws is determined by Nature; accordingly, the necessitarian approach provides a nice solution to the above problem, while the regularity approach solves the epistemological problem. Thus, I maintain that the use of laws is the best, most coherent deductive system by which we explain and manipulate the world.

Concerning the conceptions of scientific explanation and approaches to laws of nature, I see a close connection between the conceptions of explanation and the varying approaches to laws. For example, van Fraassen's pragmatics corresponds to his refusal of laws, and Salmon's causal mechanism model is closely related to his account of causation. I suggest that, comparatively speaking, Philip Kitcher's unificationist account matches the regularity approach of laws perfectly, but the accounts of D. Mellor, W. Salmon, and N. Cartwright are not so coherent. Thus, I insist that laws are essential to explanation, and epistemologically, I prefer the unificationist model of explanation and the regularity approach to laws.

Concerning causal mechanisms and lawful explanations, I provide four arguments against Salmon's causal mechanism model, claiming that laws are prior to causation and essential to explanation. I also propose that the ontological order of explanation should be from events (or processes) to causation, then to regularity (laws), and finally to the science system, but the epistemological order should be from the science system to laws to explanation and causation.

Concerning *ceteris paribus* (CP) laws, I try to answer challenges to CP laws posed by John Earman, J. Roberts, and S. Smith, and I point out that Earman's approach to laws of nature may be inconsistent with his view of CP laws. I also suggest there is some tension between Platonism and Humean accounts of laws. If we were to give up residual Platonic idealism in the contemporary sciences, it would not be difficult to accept CP laws.

Concerning explanation and reduction, I review their relation and analyze the ambitious concept of reduction into language reduction, theory reduction, discipline reduction, micro-reduction, ontological reduction, methodological reduction, epistemological reduction, explanatory reduction, etc. After doing so, I then provide brief comments on the respective types of reductionism.

Concerning scientific explanation and historical interpretation, I support Hempel's application of his explanation models to social sciences and humanities, and I respond to the objections raised by idealists. However, I do admit that interpretation cannot be covered by explanation. Thus, I claim that there should be methodological unity between the natural sciences and the human sciences. With that said, this kind of unity is not achieved by the natural sciences conquering or replacing the human sciences; instead, by mutual learning, the natural and human sciences will be unified into one "human knowledge."

As mentioned, this book tries to achieve a holistic synthesis, arguing that our scientific knowledge is a holistic system, which is constituted by laws. We use scientific laws to explain and manipulate the world, as well as to identify causation. The three concepts, explanation, law, and causation, should be understood broadly, for if they are, they can be applied in both the natural sciences and the human sciences. The hope for doing so is that boundary between nature and humanity can be opened up, and we can finally achieve a unified system of science.

The main research methods of the book include logical analysis and case studies in the history of science. In the light of the results of my research, I provide some arguments for each topic and try to achieve a Kuhnian synthesis of the topics. Hopefully, this book will provide some useful insights for related studies and will be an interesting contribution to the literature on this important topic.

The research contained herein is purely fundamental and theoretical. All three concepts, scientific explanation, laws of nature, and causation, are important issues in the philosophy of science. Indeed, at the 2008 Biennial Meeting of the Philosophy of Science Association, the problem of induction and causation (and its relation with explanation and laws) were two central topics in the general philosophy of science. Hopefully, this book can clarify these frequently used concepts in science – explanation, laws, and causation – and thus finally deepen our understanding of science. The practical value and social influence of the philosophy of science is indirect: the philosophy of science promotes the growth of science, whereas the growth of science has great practical value for and social influence on human beings.

Because explanation, laws, and causation are all frontier issues in the general philosophy of science, the research is undoubtedly difficult to undertake. Personally, I think the parts addressing explanation and laws in this book are better articulated, while the part dealing causation demands more discussion in the future. In addition, my publications associated with this project have mainly been in Chinese, although I have presented several topics in English at international conferences. I admit that I should have attempted to publish more papers in English.

This book is kindly sponsored by the Youth Project of the National Social Science Foundation (No. 04CZX004). The main ideas of most chapters were developed and articulated during my time at the University of Pittsburgh. Thus, I would like to express my gratitude to several professors at Pitt: John Earmn, John Norton, Sandra Mitchell, Merrilee Salmon, Adolf Grünbaum, Anil Gupta, and others. Professor Clark Glymour at CMU answered many of my questions about causation. Professor Malcolm Forster from the University of Wisconsin-Madison taught at Tsinghua University in the academic year of 2010–2011, and during that time, he often discussed advanced issues in the philosophy of science with me. These discussions gave me much inspiration.

I shall also thank my parents for constantly supporting my research and teaching. They have always been strong backers of my endless academic exploration. Additionally, my colleagues at the Institute of Science, Technology, and Society provide me with good conditions for conducting research. The former and current directors of the Institute, Professor Guoping Zeng and Professor Tong Wu, strongly supported me in my application to the Social Science Foundation and encouraged me to do pure and fundamental research. Professor Jiang Yang also made great contributions to the publication of this book. I also highly appreciate Tsinghua University Press, which kindly published my third book. The first two published books were *Relativism* (2003) and *Some Issues in the Philosophy of Science* (2004). The executive editor Ms. Qiaozhen Wang has worked and proofread carefully, and her tireless work has ensured the quality of the manuscript. I would also like to express my gratitude to her.

My academic achievement, if there is any of which to speak, is entirely due to my teachers' careful cultivation. In my life, I have been so fortunate to meet many teachers who excelled both in teaching skill and in moral integrity. Among those who stand out in my mind are my middle school teacher Yaobin Miao, my high school teacher Xiangchang Zhou, my MPhil supervisor Professor Shiqi Kou, and my PhD supervisor Professor Hsiu-hwang Ho, as well as Professors Yuankang Shih and Tze-wan Kwan at the Chinese University of Hong Kong. Along with those I've mentioned, there have been many other teachers in various periods of my life who also had a subtle but lasting influence on my growth. Thus, I wish to dedicate this book to my respected teachers.

1 Hempel's scientific explanation models and their problems

Section 1 Introduction

We have to explain numerous events every day. In our daily life, we may ask: What is really going on during a solar eclipse? Why do I have a cold today? What is the reason that China has developed so quickly in recent years? Similarly, we have many types of explanations in scientific research. For example, college students write experiment reports to explain why an experiment rendered certain result. We might ask, then, is there any general form of explanations? If any, what is the logical form of scientific explanation?

At the beginning of human civilization, people tended to explain nature in mythological and anthropopathic ways, attributing all natural events to humanized gods. For example, when considering the question of why it rains or thunders, the ancient Chinese believed that there are dragon kings who take care of rain and a Thunder God who creates thunder. Therefore, "agents" in myth explained natural phenomena.

After that, many philosophers tried to give the world a metaphysical explanation as they searched for the ultimate cause. For example, Aristotle used his four causes (material cause, formal cause, efficient cause, and final cause) to explain everything in the world. However, if we continue to ask where the ultimate form, effect, and telos come from, we notice that Aristotle might have had no other choice but to appeal to God as the ultimate cause, which means we cannot get rid of a "metaphysical agent."

Facing such a problem, some scientific theorists such as G.R. Kirchhoff and Ernst Mach claimed that scientists should ask *how* instead of asking *why*. The reasoning behind this is simple: to answer a *how* question, we only need a mathematical description of nature, which avoids the "metaphysical agent" problem that arises when searching for *why*.

Starting in the 1930s, circles within the philosophy of science began to focus on the general form of scientific explanations. At that time, Hans

Driesch, a German philosopher and biologist, used the term *entelechy* to explain regeneration and reproduction in biology. He argued that every creature has its own entelechy, although it is invisible and even undetectable just like an electric or magnetic field. The complexity level of entelechy increases from plants to animals. For example, the gecko can regenerate its tail after losing the former one, and a human's injured fingers can heal themselves automatically. These above examples are due to entelechy's taking effect. Hans Driesch used the term extensively to explain most biological phenomena, and he even regarded the human mind as part of entelechy.

At the 8th World Congress of Philosophy in 1934, held in Prague, Rudolf Carnap and Hans Reichenbach both criticized Driesch, saying that his explanation introduced a new terminology but didn't bring any new scientific discoveries; thus, as they argued, those explanations based on entelechy were no more than pseudo-explanations. To illustrate this idea, Carnap (1995) wrote a chapter specifically to discuss the general form of scientific explanation (pp. 12–19).

Afterwards, Karl Popper and Carl Gustav Hempel both had discussions about scientific explanation, but it is commonly believed that Hempel's discussion is conducted more clearly and completely. Because of this, let us begin with Hempel's scientific explanation models.

Section 2 Hempel and his contributions

Carl Gustave Hempel, also called Peter by his friends, was born in Oranienburg, a town near Berlin, on January 8th, 1905. Hempel was well educated in his early years, going to the University of Göttingen in 1923 to study mathematics under David Hilbert and Edmund Landau, while also learning symbolic logic from Heinrich Behmann. In the same year, Hempel went to Heidelberg University to study mathematics, physics, and philosophy. Under the guidance of Hans Reichenbach, Hempel began his doctoral studies at the University of Berlin in 1924. While studying in Berlin, he also followed Max Planck and John von Neumann to learn physics and logic.

Reichenbach's two books *Pseudoproblems in Philosophy* and *Logical Structure of the World* deeply inspired Hempel, and, encouraged by Reichenbach, Hempel decided to visit the University of Vienna for one year. While in Vienna, he studied with Moritz Schlick, Rudolf Carnap, and Friedrich Waismann. He was also able to engage in academic discussions with Otto Neurath, Herbert Feigl, Hans Hahn, and even with Ludwig Wittgenstein.

Due to the rise of Nazism in Germany, Reichenbach left Berlin in 1933, and Hempel chose Wolfgang Koehler and Nicholi Hartman as his new advisors, finally receiving his doctorate degree in 1934. Although he was not Jewish, he detested the dominance of the Nazis in Germany, so he seized an opportunity to immigrate to Brussels, Belgium, that same year. While there, Hempel became acquainted with Paul Oppenheim, an independent German scholar and chemistry industrialist. They did a great deal of work together, such as the well-known thesis "Studies in the Logic of Explanation," which made scientific explanation become one of the central issues in the philosophy of science.

For the academic year 1937–1938, Carnap invited Hempel to visit the University of Chicago as his research assistant. After this, Hempel finally immigrated to the United States and taught summer courses at the City University of New York. In 1940, Hempel moved to Queen's College where he stayed until 1948. From 1948 to 1955, he was an assistant professor at Yale University and published *Fundamentals of Concept Formation in Empirical Science*. In 1955, he was named the Stuart Professor of Philosophy at Princeton University, where he later published the masterpieces *Aspects of Scientific Explanation* (1965) and *Philosophy of Natural Science* (1966). From 1976 to 1985, Hempel spent his time teaching and researching at the University of Pittsburgh as a University Professor. Finally, after his retirement, Hempel lived his last years in Princeton, where he passed away on November 9th, 1997 (Fetzer, 2000, pp. xv–xxvi).

Section 3   Models of scientific explanation

1. The DN model of scientific explanation

Hempel put forward the *Deductive-Nomological model* of scientific explanation, which is also called the *DN model*. The structure of the DN model can be shown as follows:

$C_1, C_2, \ldots, C_k$ (Initial conditions)
$L_1, L_2, \ldots, L_r$ (General laws)

E (Description of empirical phenomenon to be explained)

In this model, C is the initial condition, while L is general law, with both constituting the explanans. We can deduce E from the conjunction of L and C, which means the explanandum E is a logical consequence of the explanans.

Hempel provides an example about a frozen crack on a car radiator. The initial conditions are as follows:

(a) The car has been outside for the whole night.
(b) The outside temperature was below 25°F, and atmospheric pressure was normal.

(c) The most pressure that the car radiator could bear is P_0.
(d) The radiator was full of water and sealed up.

General laws include:

(a) The freezing point of water is 32°F under normal atmospheric pressure.
(b) When the temperature is below freezing point and the volume is unchanged, the pressure of water will rise as the temperature goes down. There is thus a function relationship between temperature and pressure.

From these initial conditions and general laws, we can calculate the pressure P of water stress on the radiator, which is larger than the maximum pressure that the radiator can bear, P_0. Therefore, we can logically deduce that the radiator cracked, which is the empirical phenomenon to be explained.

Hempel claims that the model of DN must satisfy three logical conditions and one empirical condition. Logical conditions are as follows:

(a) Explanandum must be the logical consequence of explanans. In other words, the explanandum must be logically derived from information of the explanans; otherwise, the explanans is not enough to explain the explanandum. The aim of the condition is to ensure the correlation between explanans and explanandum is inevitable, not accidental. If we can logically reach explanandum from explanans, explanandum must be true when explanans is true. This condition is also called the deductive thesis.
(b) Explanans must include general laws, and those laws are necessary when deriving explanandum. Inclusion of general laws is to ensure that the derivation of explanandum from explanans is replicable and also regular. This condition is also called the covering law thesis. Certainly, explanans usually also includes descriptions, which are initial conditions and not lawlike.
(c) Explanans must include empirical contents, which means that at least the explanation can be tested by experiments or observations in principle. Therefore, Driesch's *entelechy* is excluded from scientific explanations of life phenomenon since entelechy cannot be tested through experimentation or observation.

The DN model needs to satisfy an empirical condition as well: statements consisting of explanans must all be true. If general laws or initial conditions are false, it is not a scientific explanation even if we can derive the explanandum from the explanans.

2. The IS model of scientific explanation

Dealing with statistical explanations in scientific research, in 1962 Hempel devised the *Inductive-Statistical model* (or *IS model*), based on the DN model. The structure of the IS model is as follows:

F_i	Initial conditions
$p(O, F) = r$ (r is close to 1)	Statistical law
O_i	Phenomenon to be explained

For example, take the condition "I feel a cold breeze after sweating" as an initial condition. People who feel a chill after sweating don't always get a cold, but they do have a higher chance (maybe 80% or higher) of being attacked by a cold. Therefore, the law we have is a statistical one: catching a chill after sweating means an 80% possibility of catching a cold. The conjunction of initial condition and statistical law support the explanandum "I caught a cold" to a high degree, so it is a scientific explanation.

It is noteworthy here that we can logically derive "I have an 80% possibility of catching a cold" from initial conditions and statistical laws. For this sort of inference, Hempel coined the term *Deductive-Statistical model* (*DS model* for short) (Hempel, 1965, pp. 380–381). The logical form of DS model is as follows:

F_i	Initial conditions
$p(O, F) = r$ (r is close to 1)	Statistical law
$p(O_i) = r$	Phenomenon to be explained

However, the DS model only shows us the possibility of a specific event, such as "I have an 80% possibility of catching a cold," instead of a certain event such as "I caught a cold." Therefore, Hempel paid more attention to the DN model and the IS model.

Based on the logical form of the IS model, we can only logically derive "I have an 80% possibility of catching a cold" without reaching the explanandum "I caught a cold." Therefore, with the IS model, explanans gives explanandum a high level of support, but the explanandum is not inevitable. The inference here is inductive, not deductive. Accordingly, there are two lines between explanans and explanandum in the IS model in order to show the differences when contrasted with deductive derivation of the DN model (one line between explanans and explanandum).

The IS model must satisfy three logical conditions and two empirical conditions. The logical conditions are as follows:

(a) Explanandum must have a high possibility of being derived from explanans.
(b) Explanans must include at least one statistical law that is necessary for derivation of explanandum.
(c) Explanans must include empirical content, which means, at least in principle, that the explanation can be tested through experimentation or observation.

To continue, the empirical conditions are the following:

(d) Statements of explanans must be true.
(e) Statistical laws of explanans must satisfy the requirement of maximal specificity (RMS for short).

The conditions (a)–(d) are basically the same as the conditions of the DN model. The fifth condition of the IS model is to choose the maximal specific sample. For example, Jack passed out after eating a pound of candies. If we used the statistical law that people may pass out after eating a pound of candies with a possibility lower than 1 in 10,000, this is too little for a satisfactory explanation. However, if Jack is diagnosed with diabetes and there is a law that people suffering from diabetes have a 99% probability of passing out after eating a pound of candy, then we have the maximal specificity of Jack's case and have therefore explained his passing out successfully.

Hempel proposed both the DN model and the IS model of scientific explanation, yet there are some notable differences between the two models:

(a) Laws of the DN model are general deterministic laws, while the IS model usually uses statistical laws.
(b) The inference of the DN model is deductive so that its results are inevitable, logical derivations. The inference of the IS model is inductive, which means the reliability of inferences is determined by the probabilities of laws. Even if the explanans is true, the explanandum may not happen as well.

On the other hand, Hempel points out that the DN and the IS models share a similar form: both include scientific laws, which is essential to explanations. Scientific explanations must include scientific laws, so the requirement is called the covering law thesis. Hempel (1965) also calls his models *covering law models* (p. 412).

3. Supplementary specification of scientific explanation

The covering law thesis is the essential requirement of Hempel's scientific explanation models. But what are scientific laws? Hempel tries to distinguish scientific laws from accidental generalizations. For example, the statement "all metals are electrically conductive" is a scientific law. In contrast, "all coins in my pocket are made of nickel" is a true, universal statement, but we never regard it as a scientific law.

If we cannot distinguish scientific laws from accidental generalizations, some explanations will be *ad hoc*. For example, to explain why the pen in my pocket is electrically conductive, we can use the law "all metals are electrically conductive" or say "all things in my pocket are electrically conductive." Obviously, the latter is not a scientific explanation.

Hempel tries to summarize the general characterization of lawlike sentences so that scientific laws are those descriptions that are both lawlike and true, but soon he finds that we cannot summarize the logical form of lawlike sentences. For example, no one has ever seen a gold block heavier than 1 million tons, so we can make the generalization that "all gold blocks are lighter than 1 million tons." Is this proposition a scientific law? The answer is based on our understanding of the world. If in the future scientists find that gold decomposes once its weight reaches 1 million tons, then the generalization can be considered a scientific law. Otherwise, it is an accidental generalization. Therefore, what is considered scientific law relies heavily on scientific research and is not only determined by logical analysis.

In addition, there is a "problem of ambiguity" when it comes to statistical explanation. For example, most of us (99% or more) will live more than 5 years after turning 30 years old. According to this statistical law, we can conclude that Jack will live at least 5 years after his 30th birthday. However, patients with terminal lung cancer have a high chance (96% or more) of dying within 5 years. Based on this new statistical law, we can make the inference that Jack, who is patient with terminal lung cancer, won't live for more than 5 years. So, after all those inferences, how long can Jack live? Different explanations give us different results, which shows the problem of ambiguity of statistical explanation. The logical form of the problem is as follows:

Argument 1   Argument 2

$P(G/F) = r$ $P(\neg G/H) = r'$

Fa Ha

$$======= [r] \qquad ======= [r']$$

Ga $\neg Ga$

Hempel argues that statistical explanations rely on our background knowledge (such as whether or not Jack is a terminal lung cancer patient) and therefore are not as objective as the DN model. He calls this "epistemic relativity." In order to avoid the problem of ambiguity, we need to introduce a "requirement of total evidence" into statistical explanation, which means all evidence about the explanandum should be demonstrated.

4. Variations of scientific explanations

Hempel reminds us that, even in the natural sciences, not all explanations can completely fit into his DN, IS, or DS models. He suggests there are numerous variations of explanation models in scientific practice for convenience or other reasons, and then he proposes three particular forms: elliptic explanation, partial explanation, and explanation sketch (Hempel, 1968, pp. 62–64).

Elliptic explanation is to omit laws and initial conditions that everyone knows in order to form a simplified explanation. If we were to add those omitted initial conditions and laws, the complete explanation would still fit into the DN or IS model. For example, when we try to explain why copper is electrically conductive, sometimes we just say "because copper is metal." The explanation omits the law that "all metals are electrically conductive," which is a well-known law for everyone. If the law is added into the explanation, the complete explanation will be: "All metals are electrically conductive, and copper is a metal, so copper is electrically conductive," which apparently satisfies the DN model. Indeed, sometimes we use "all metals are electrically conductive" to explain why copper is a conductor, omitting that fact that "copper is a metal" from the explanation.

In a partial explanation, the explanandum is just a subset of the conclusion, which is derived from the explanans. For example, according to some psychological law, people tend to lose things when they are extremely depressed. However, we are not able to explain precisely or predict what item will actually go missing. From the initial condition "Jack is extremely depressed" and the relevant psychological law, we can only explain "Jack lost something" instead of "Jack lost his wallet." However, "Jack lost his wallet" is a subset of "Jack lost something," which thus constitutes a partial explanation.

In some explanations, laws and initial conditions are too complicated to be stated exactly, so we can only give an outline of an explanation or a sketch for the explanandum. An explanation sketch is different from pseudo-explanation because it requires researchers to do more empirical work to fill out its content. This is because an explanation sketch offers an empirical hypothesis, which can be verified or falsified in principle.

For example, the global financial crisis in 2008 involved many factors, and the relevant laws of economics were very complicated and difficult to describe. However, we can try to explain it in the following way: "The United States had a subprime mortgage crisis in 2007; the economy of the United States has great influence on the global economy; it finally resulted in the global financial crisis in 2008." This offers a brief explanation outline or sketch for the global financial crisis, but as it demonstrates, although explanation sketches are apparently useful, they often lack details.

Section 4   Problems of scientific explanation

Hempel built up brilliant models of scientific explanation, and in doing so, he made great contributions to science. Still, no one is wise at all times, and philosophers have found many problems with Hempel's models of scientific explanation.

1. *Explanation and prediction*

Hempel (1965) argues for the structural identity of explanation and prediction:

(a) Every adequate explanation is potentially a prediction.
(b) Every adequate prediction is potentially an explanation. (p. 367)

Since scientific explanation is an argument, this means the explanandum can be deduced or induced from the explanans. If the explanandum is known, the argument is an explanation; if the explanandum is unknown, the argument is a prediction.

Nevertheless, many philosophers criticize the structural identity of explanation and prediction. Michael Scriven, for example, gives the argument of a syphilitic mayor: imagine that Jones, the mayor of a small city, suffers from some sort of paresis. Since the paresis usually is caused by syphilis, and Jones has been a syphilitic for many years, his longtime syphilis thus explains his paresis. However, the probability of paresis among untreated syphilitics is low (roughly 10%, meaning about 90% of syphilitics are not susceptible to paresis). With this in mind, if we use the statistical syllogism, the prediction shall be "Jones won't get paresis," so this argument of the syphilitic mayor shows that explanation and prediction have different results and that there is no structural identity (Scriven, 1959).

Scriven and other philosophers also point out that the theory of evolution can explain, but cannot predict, species variation. For example, we can explain why leopards run so fast and why giraffes have such long necks with the theory of evolution: these are results of natural selection. However, the theory of evolution cannot tell us what future species will look like. Therefore, the theory of evolution can explain but cannot predict (Scriven, 1962).

In addition, Scriven claims that some events can only be explained afterwards but cannot be predicted beforehand. For example, we usually investigate the causes of an accident to explain a bridge collapse after it happens. We cannot predict the bridge collapse in advance. Aviation accidents often result in heavy casualties, so if we could predict accidents in advance, tragedies would be avoided efficiently. However, what we can do in reality is simply give conclusive explanations for disasters. In such events, explanations and predictions are not identical (Scriven, 1963).

With that said, Hempel does provide some counterarguments to these criticisms. Concerning the argument of the syphilitic mayor, Hempel believes that since the probability of paresis among untreated syphilitics is low, "syphilis causing paresis" is not a good IS explanation (a good IS explanation has to satisfy the high-probability requirement). He also believes that Darwinism offers a partial and statistical explanation of the evolution of species. Since mutations and environmental factors to a great extent are random and complex, biology cannot predict specific new species, but predictions and explanations are still identical. Concerning explanations after accidents, Hempel argues that if we could have known every specific initial condition in advance, accidents would be not only explicable, but also predictable.

2. Asymmetry Thesis

Structural identity of explanations and predictions sometimes is also called the symmetry thesis. However, numerous philosophers of science have pointed out that many scientific explanations are asymmetrical – that is, although event A and B are regularly connected, A can explain B, while B cannot explain A.

For example, according to the principles of geometry and of optics, the length of a flagpole has a proportional relation with the length of its shadow at a given time of day. Therefore, we can calculate the length of its shadow by measuring the length of the flagpole, which also can be inferred from the length of the shadow. However, we usually believe that the length of the flagpole can explain the length of its shadow, but not vice versa. Therefore, an explanation is not always symmetrical.

Similarly, according to classical mechanics, there is correlation between the length of a single pendulum and its period:

$$T = 2\pi\sqrt{l/g}$$

We can calculate the period according to its length, but also calculate the length according to its period:

$$l = gT^2/4\pi^2$$

However, we usually say that the length of a pendulum explains its period; few of us believe that the period of a pendulum explains its length.

Asymmetry is not just temporal: some events that happen in advance cannot fully explain the events that happen afterwards either. For example, drastic changes in barometric readings usually indicate storms in the future, but we would never think that changes in barometric readings can success-fully and sufficiently explain the storms coming afterwards.

The asymmetry thesis is well connected with the structural identity of prediction and explanation. Therefore, all examples described above can also be used to demonstrate differences between explanation and prediction.

3. The irrelevance objection

David-Hillel Ruben, a philosopher of social science at the University of London, has proposed the irrelevant objection to Hempel's models of sci-entific explanation. Here is one example he borrows from Ardon Lyon (Ruben, 1990, p. 182):

(a) All metals are electrical conductors.
(b) All electrical conductors are subject to gravitational attraction.

(c) All metals are subject to gravitational attraction.

Although Lyon's example satisfies all the requirements of the DN model, it is not an explanation of "all metals are subject to gravitational attraction" because gravitational mass is the true reason, and electrically conductivity has nothing to do with gravitational attraction. Thus, electrical conductiv-ity is irrelevant information, which Hempel's explanation models cannot effectively exclude.

Peter Achinstein (1983) gives another example (pp. 168–171):

(a) Jones ate a pound of arsenic at time t.
(b) Anyone who eats a pound of arsenic will die within 24 hours.

(c) Jones died within 24 hours after t.

This example satisfies the DN model as well; however, Jones didn't actually die of arsenic poisoning. He was so unfortunate that he died of a car accident soon after eating arsenic, which means that eating arsenic was irrelevant to his death. Hempel's explanation models cannot successfully deal with such an occasion.

Timothy McCarthy (1977) also proposes a formula to criticize Hempel (pp. 159–166):

$$\forall x(Ax \rightarrow Bx)$$
$$C(e) \wedge A(o)$$
$$\neg B(o) \vee \neg C(e) \vee D(e)$$
$$\overline{}$$
$$D(e)$$

Since the first of the explanans can be a universal law, and the conjunction of explanans can lead us to logically derive the explanandum, McCarthy's formula satisfies the DN model in every aspect. However, it may not be a good scientific explanation. Considering the following example, such an "explanation" is, in fact, seen to be quite ridiculous:

All metals conduct electricity.
The forest was stuck by lightning, and this watch is made of metal.
This watch does not conduct electricity, or the forest was not struck
    by lightning, or the forest caught fire.

The forest caught fire.

All of these examples constitute irrelevant objections to Hempel's models of scientific explanation.

4. Requirement of maximal specificity

Hempel also suggests the requirement of maximal specificity (RMS) for the IS model, but RMS is questioned by Wesley Salmon. Consider that salt has a high probability (say 95%) of dissolving in cold water within 5 minutes. We can say some "dissolving spells" to salt, making them into "hexed salt," and then we can have a lawlike generalization: hexed salt has a high probability of dissolving in cold water within 5 minutes.

Now we need to explain a new event: namely, that some hexed salt dissolved in water. According to RMS, we should make it specific that the salt is hexed salt, thus the explanation is as follows:

Hexed salt has a high probability of dissolving in cold water within
    5 minutes.
Hexed salt was put into water.

Hexed salt dissolved within 5 minutes.

Of course, this is not a satisfactory scientific explanation. Therefore, Salmon
claims that Hempel's RMS should be corrected by the requirement of the
maximal class of maximal specificity. Since both hexed and normal salt
have the characteristic of dissolving within 5 minutes, we should choose the
maximal class of maximal specificity, "salt," instead of the smaller class,
"hexed salt."

However, Salmon also argues that this correction is no help, for we can
hardly choose the proper "maximal class of maximal specificity." For example, salt and baking soda both have a high probability of dissolving in cold
water within 5 minutes. Consequently, when we explain the dissolution of
hexed salt, should we choose "salt and baking soda" as the maximal class of
maximal specificity? To do so would be to explain the dissolution of hexed
salt with the statement "salt and baking soda both have a high chance of
dissolving in cold water within 5 minutes." This is certainly inconsistent
with our scientific intuition. Thus, because Hempel's explanation models
have been met with a host of questions and challenges, many philosophers
of science have attempted to create new conceptions and approaches for
scientific explanation.

2 Six decades of scientific explanation

Hempel first proposed the logic of scientific explanation with Paul Oppenheim in 1948. Forty years later, W. Salmon reviewed the academic development of scientific explanation from 1948 to 1989 in his book *Four Decades of Scientific Explanation*. Today, the study of scientific explanation has been developing for more than 60 years. During these past 60 years, there has been much criticism and development of Hempel's explanation models. Here, only some of the more important and influential views will be introduced.

Section 1   Bas C. van Fraassen: pragmatics of scientific explanation

Hempel emphasizes the logic of explanation, which usually gives no regard to the context. In other words, if the DN model and IS models hold, then they can hold in any context. Yet, at the same time, van Fraassen's pragmatics of explanation is focused on the context of explanation. Because of this, the differing explanation models are worth comparing.

Bas C. van Fraassen was born in the Netherlands on April 5, 1941, and he immigrated to Canada in 1956. After he received his bachelor's degree at the University of Alberta in 1963, he moved to the United States, where he studied with Wilfrid Sellars and Adolf Grünbaum at the University of Pittsburgh. Throughout his career, he has taught at Yale University, the University of Toronto, the University of Southern California, and Princeton University, and he is now teaching at San Francisco State University. His main publications include *An Introduction to the Philosophy of Time and Space* (1970), *The Scientific Image* (1980), *Laws and Symmetry* (1989), *Quantum Mechanics: An Empiristic View* (1991), *The Empirical Stance* (2002), and *Scientific Representation: Paradoxes of Perspective* (2008).

Van Fraassen (1977) holds that traditional explanation models present three ideas: (1) explanation is a relation simply between a theory or hypothesis and the phenomena or facts; (2) explanatory power cannot be logically separated from certain other virtues of a theory, notably truth or acceptability; and (3) explanation is the overriding virtue, the end of scientific inquiry (pp. 143–150).

Van Fraassen argues against the idea that scientific explanation, truth, and acceptance of a theory are equivalent. In his opinion, scientists accept a scientific theory simply because the theory is empirically adequate – that is, to save the phenomena. Therefore, the acceptance of a scientific theory does not need to rely on an acceptance of its truth, let alone an explanation of all phenomena in the area. Scientific explanation is neither the overriding virtue nor the end of scientific inquiry. For instance, if we have to explain the probability phenomena in quantum mechanics, we might need to introduce hidden variables; however, doing so will also bring with it metaphysical baggage.

Van Fraassen thinks that the traditional explanation models have two prejudices. (1) Philosophy of science must indicate the sufficient conditions and necessary conditions concerning why theory T explains phenomenon E. (2) Explanatory power is a virtue of a theory itself (or its relation with the world), such as simplicity, truth, empirical adequacy, etc. Yet, his own suggestion is this:

> So scientific explanation is not a (pure) science but an application of science. It is a use of science to satisfy certain of our desires; and these desires are quite specific in a specific context, but they are always desires for descriptive information.
>
> (van Fraassen, 1980, p. 156)

Therefore, a successful scientific explanation is usually an appropriately successful description with information, and the truth and acceptance of a scientific theory are irrelevant.

In van Fraassen's (1980) view, an explanation is an answer to a *why-question*; therefore, a theory of explanation has to be a theory of a why-question (p. 134).

Sylvain Bromberger (1966), a professor at MIT, originally studied the why-question (pp. 86–108), and van Fraassen later developed the research further. A why-question always begins with a *why*, for instance, "Why did Adam eat the apple?" However, in a different context, the meaning of a why-question may be different. For instance, "Why did Adam eat an apple?" can have the following three meanings in different contexts:

(1) Why did Adam (not other people) eat an apple?
(2) Why did Adam eat an apple (not another fruit)?
(3) Why did Adam eat (not play with) an apple?

Therefore, a why-question should include not only a topic (the meaning, P_k, represented by the question itself), but also a contrast-class X. A contrast-class indicates why P happened, not other cases in the contrast-class X. For instance, when considering the question "Why did Adam eat the apple?" if what we care about is why Adam ate an apple instead of another fruit, its contrast-class in the context shall be:

(1) Adam ate a banana.
(2) Adam ate a pearl.
(3) Adam ate an orange. . .

Furthermore, this topic and its contrast-class also have a "relevance relation." For instance, "Adam's favorite fruit is apple" and "Adam ate an apple (not a banana or an orange)" have a relevance relation, but they have no relevance relation with the statement "The solar system has 9 or 8 planets."

To summarize, the why-question Q expressed by an interrogative in a given context is determined by three factors:

(1) The *topic, P_k*,
(2) The *contrast-class $X = \{P_1, \ldots, P_k, \ldots\}$*
(3) The *relevance relation, R*

Preliminarily, we may identify the abstract why-question with the triple consisting of these three:

$$Q = <P_k, X, R>$$

With this, a proposition A is considered relevant to Q exactly if A bears relation R to the couple $<P_k, X>$.

We must now define what the direct answers to this question are. To begin, let us inspect the form of words that will express such an answer: (*) P_k in contrast to (the rest of) X because A.

Therefore, the why-question *presupposes* exactly that (a) its topic is true; (b) in its contrast-class, only its topic is true; and (c) at least one of the propositions that bears its relevance relation to its topic and contrast-class is also true. B is a *direct answer* to question $Q = <P_k, X, R>$ exactly if there is some proposition A such that A bears relation R to $<P_k, X>$, and B is the proposition that is true exactly if (P_k; *and* for all $i \neq k$, not P_i ; and A) is true (van Fraassen, 1980, pp. 143–145).

For instance, "Adam's favorite fruit is apple" makes clear that P_k (Adam ate an apple.) happened, while other cases in contrast-class X (such as "Adam ate a banana" and "Adam ate an orange") did not happen. So, it bears the *relevance relation R* with the topic P_k and *contrast-class X*. "Adam's favorite fruit is apple" answers the question "Why did Adam eat an apple?" in this context and is therefore a successful explanation.

Thus, van Fraassen's critique is that the discussion of scientific explanation is wrong from the beginning. Traditional opinion holds that scientific explanation describes the relation between theory and fact, but it is actually the relation between theory, fact, and context. A scientific explanation is an answer to a question and the demand for needed, relevant information. The question "Why did P happen?" has different meanings in different contexts. Therefore, the answer or explanation shall be different accordingly.

After clarifying the pragmatics of scientific explanation, van Fraassen (1977) writes from a pragmatic viewpoint: "Explanation is indeed a virtue; but still, less a virtue than an anthropocentric pleasure" (pp. 143–145).

Section 2   Salmon: causality and explanation

Wesley C. Salmon was a well-known American philosopher of science and logician. At UCLA, under Hans Reichenbach, Salmon received his PhD degree in philosophy in 1950. He taught at UCLA, Washington State University, Northwestern University, Brown University, Indiana University, and the University of Arizona. Salmon left the University of Arizona to join the Department of Philosophy at the University of Pittsburgh in 1981, and he finally retired in 1999. His main publications include *Logic* (1963), *The Foundations of Scientific Inference* (1967), *Statistical Explanation and Statistical Relevance* (1975), *Scientific Explanation and the Causal Structure of the World* (1984), *Four Decades of Scientific Explanation* (1990), and *Causality and Explanation* (1998).

In his writings, Salmon criticizes Hempel's explanation models from the relevance perspective. For instance, we can construct a counter-example of the DN explanation model as follows:

Every man who regularly takes birth control pills avoids pregnancy.
John took his wife's birth control pills regularly during the past year.

John avoided becoming pregnant.

This explanation fits the DN explanation model well, but we would never believe that it is a good explanation. The information concerning whether John took birth control pills is irrelevant because a man can never be pregnant. This example shows that relevance is an essential factor in scientific explanation.

Salmon also criticizes the high probability requirement of Hempel's IS explanation model and proposes that statistical relevance, rather than high probability, is the key to statistical explanation. He gives the following example:

> Most people who have a *neurotic* symptom of *type N* and who undergo psychotherapy experience relief from that symptom.
> Jones had N-type neurosis and accepted psychotherapy treatment.
> ── (r)
> Jones' N-type neurosis has been recovered.

According to Hempel's IS model, if there is a high probability of r, then this is a good statistical explanation. But since a certain proportion (r') of neurotic patients will automatically recovery without any treatment, Jones' recovery is not necessarily due to the treatment, but potentially due to natural recovery. Therefore, whether the psychotherapy treatment is the relevant key of the explanation, is not determined by the strength of the probability r, but by whether r is greater than r' – that is to say, whether psychotherapy treatments increase the proportion of recoveries.

Salmon proposes statistical relevance as a kind of probability calculation. Its definition is the following: given the condition A, the factor C is statistically relevant to the factor B if and only if $P(B/A\&C) \neq P(B/A)$. For instance, if the probability of Jones' recovery can be changed by psychotherapy treatment, then it is statistically relevant. If not, then the psychotherapy treatment shall be statistically irrelevant.

Finally, Salmon (1998) proposes five conclusions. (1) We must put the "cause" back into "because." Although some explanations are not causal, scientific explanation must demonstrate a cause. Due to the asymmetry of causation and time, scientific explanation is asymmetric too: the cause explains the effect; an earlier event can explain a later event, but not vice versa. (2) A high probability is neither the sufficient condition nor the necessary condition for scientific explanation. The key is whether the explanans is statistically relevant, that is, the explanans can increase the probability of the appearance of the explanandum. (3) Hempel's "principle of essential epistemic relativity" of the IS explanation model must be given up. Salmon introduces the concept of an "objectively homogeneous reference class" and suggests that a statistical explanation has the same objective of correctness as the DN model. (4) The theory of scientific explanation should include the pragmatics of explanation. (5) We should not pursue the formal model of scientific explanation, which is universally applicable to all sciences; instead, we should observe the real form of explanation in specific sciences (pp. 302–319).

However, Salmon also acknowledges that there are three controversies in his account of scientific explanation. (1) Some trouble arises when discussing the nature of laws of nature because there is no universally acceptable criterion to distinguish between scientific laws and accidental generalizations. (2) Can there be any statistical explanation of specific events? Some philosophers think explanation can only be deductive, so we can only logically deduce statistical description by statistical laws, but we can't statistically explain a certain event. That is, they only accept the DS model, not the IS model. We can only explain "how the world works" but not "what happens." (3) Philip Kitcher distinguishes between two approaches to scientific explanation: "bottom up" and "top down," which can be also called "local" and "global." Both Hempel's and Salmon's explanation models are bottom up or local, while Kitcher prefers the top-down approach. Either way, Salmon believes that this is still an unsolved issue of scientific explanation.

Section 3   Explanation: global and local

It was Michael Friedman who first drew a distinction between the global and local forms of explanation. He received his PhD from Princeton University in 1973 and is the director of Patrick Suppes Center for History and Philosophy of Science at Stanford University. His areas of interest include Kant, the philosophy of science, and the history of 20th century philosophy. His publications include *Foundations of Space-Time Theories* (1983), *Kant and the Exact Sciences* (1992), *Reconsidering Logical Positivism* (1999), *A Parting of the Ways: Carnap, Cassirer, and Heidegger* (2000), and *Dynamics of Reason* (2001).

Friedman proposes that scientific explanation is a global, rather than local, understanding provided by science, which is represented in the simplification and unification of our world image. He writes:

> On the view of explanation that I am proposing, the kind of understanding provided by science is global rather than local. Scientific explanations do not confer intelligibility on individual phenomena by showing them to be somehow natural, necessary, familiar, or inevitable. However, our over-all understanding of the world is increased.
>
> (Friedman, 1971, pp. 5–19)

Philip Kitcher further distinguishes bottom-up and top-down approaches to scientific explanation. Kitcher was born in England and received his PhD in History and Philosophy of Science from Princeton University in 1974. Now he is the John Dewey Professor of Philosophy at Colombia University.

His areas of specification include the philosophy of science, pragmatism, naturalistic ethics, and philosophy in literature. His publications include: *Abusing Science* (1982), *The Nature of Mathematical Knowledge* (1983), *Vaulting Ambition* (1985), *The Advancement of Science* (1993), *The Lives to Come* (2001), *Science, Truth, and Democracy* (2001), *In Mendel's Mirror* (2003), and *Finding an Ending* (2004).

Kitcher proposes that the bottom-up approach is a kind of local explanation, which tries to derive superficial phenomena from fundamental scientific laws, while the top-down approach is a kind of global explanation, which tries to provide a global understanding for phenomena.

Kitcher's criticism is that traditional scientific explanation models are all local, and he advocates that scientific explanation should be global: to explain something is to put it into the global pattern. For this reason, he raises the concept of "explanatory store." Given every phenomenon K to be explained, if $E(K)$ is the most likely demonstration set to unify K, then $E(K)$ constitutes the explanatory store for K.

Kitcher takes Newtonian mechanics as an example in the history of science, as there are many phenomena – such as the planetary motion, tides, free-falling objects, etc. – that demand explanation. Since the law of universal gravitation and Newton's three laws of motion constitute a *schematic sentence*, the mathematical methods of Newtonian mechanics also provide an *argument pattern*. Thus, these complex phenomena can be unified by Newtonian mechanics. In this sense, Newtonian mechanics is a system that constitutes the explanatory store of movement phenomena.

Kitcher thinks that his "global explanation" can successfully avoid the asymmetry problem, irrelevance objection, and the "genuine law-accidental generalization distinction" in the traditional explanation models. The asymmetry problem is due to the fact that some scientific laws have logical equivalence; hence the effect can be derived from the cause and vice versa. But the global explanation searches for the bottom-up global understanding of empirical phenomena, which has direction; therefore, it can avoid the asymmetry problem.

His global explanation can also exclude the irrelevance objection. Since the global explanation provides a kind of argument pattern, even if we use the irrelevant information to explain the phenomena only once, the argument pattern probably cannot be used later; therefore, the final explanatory information must be relevant.

Because Hempel failed to distinguish scientific laws and accidental generalizations, Kitcher's answer to the problem is that laws or casual relations are determined by their place in the simplest but broadest system of scientific theory. Those that can be found in the global pattern are scientific laws. Otherwise, it is merely a case of accidental generalization. Kitcher (1981) finally uses one slogan to conclude his scientific explanation model: "only connect" (pp. 507–531).

Section 4 The DNP model of scientific explanation

Peter Railton received his PhD in 1980 at Princeton University, where he chose scientific explanation as his research project, and his PhD dissertation was titled *Explaining Explanation*. His later major research has focused on ethics, especially metaethics. Now he is a professor at the University of Michigan and a fellow of the American Academy of Arts and Sciences.

Railton proposes the *Deductive-Nomological Model of Probabilistic Explanation* (or *DNP model*). The core idea of Railton's DNP model is that scientific explanation must clarify the internal mechanism: to seek explanation is to seek an internal mechanism.

In his opinion, if the world is a huge machine, then the purpose of scientific explanation is to search for its structure and working mechanism, which shall be more important than prediction or controlling effects. He proposes the DNP model in the following form (Railton, 1978, pp. 206–226):

(a) To derive the statistical law (b) from scientific theories.	theoretical derivation
(b) $\forall t \forall x [Fx,t \rightarrow P(Gx,t)] = r$	statistical law
(c) Fe,t_0	initial conditions
————————————————	deductive reasoning
(d) $P(Ge,t_0) = r$	event propensity
(e) Ge,t_0	parenthetic addendum

For instance, we want to explain why, at t_0, e has property G, which can be represented as Ge,t_0. First, we derive the statistical law (b), and then we substitute the initial condition (c) of c; having done so, we can logically deduce the probability that Ge,t_0 will happen (not that Ge,t_0 actually happened). If we want to explain why Ge,t_0 actually happened, then we need a further step: the parenthetic addendum (e).

For example, the decay of uranium-238 emitted an α particle at a certain time. To explain the phenomenon, we need to calculate the probability of decay according to the half-life period of uranium 238: $1-\exp(-\lambda_{238} \times \theta)$. Then we calculate the event propensity, or the decay probability r of uranium 238 and the probability that it emits α particles at that time: $r = 1-\exp(-\lambda_{238} \times \theta)$. Regardless of whether the probability is high or low, because we have already indicated the internal mechanism of the decay of uranium, we can explain the phenomenon with the parenthetic addendum.

However, according to Hempel's IS model, given the statistical law and the initial conditions, the probability that the explanans supports the explanandum (uranium-238 decayed and emitted an α particle) is r. If r is close to 1, then the explanation holds, but if r is close to 0, then uranium-238 will not decay.

Railton's criticism is that the inductive inference of the IS model doesn't necessarily explain a low-probability event. But if the DNP model can illustrate the internal mechanism of the event (e.g., calculate the probability of uranium-238 decay and emitting an α particle), then even a low-probability event can be explained very well.

Railton also disagrees with Hempel's thesis of epistemic relativity and the requirement of maximal specificity of statistical explanations. For example, if 23% of uranium-238 emits 4.13 MeV α particles, while 77% of uranium-238 emits 4.18 MeV α particles, how can we then explain that uranium-238 emits a 4.13MeV α particle at a certain moment?

Hempel's IS explanation depends on our relative knowledge; whether or not we know a certain portion of uranium-238 will emit different α particles with different energy will inevitably influence our explanation of the event. Therefore, the IS explanation has some degree of subjectivity and is therefore relative, but Railton thinks that his DNP explanation can avoid this problem. Because we can find the internal mechanism of emitting α particles, we can accurately calculate the propensity of uranium-238 to emit 4.18 MeV α particles, and finally we can achieve a complete and objective explanation.

To sum up, Railton's DNP model has five features: (1) all explanations are objective; (2) to explain is to provide relevant information; (3) explanation requires not only laws, but also an account of internal mechanisms; (4) real probability explanations need probability laws, which presupposes indeterminism; and (5) probability explanations do not need the requirement of high probability.

Railton's DNP model provides the propensity interpretation of quantum mechanics: probability represents the physical propensity of a single chance system producing a certain result; it can be used in the single event and has casual responsibility for the final result. We need to calculate the probability of a certain event happening by using quantum mechanics. Thus, propensity can explain the internal mechanism of an event.

Some philosophers criticize this interpretation, saying that the application of the DNP model is very limited and ignores many probability events of everyday life in non-quantum mechanics circumstances, such as gambling, classical thermodynamics, insurance actuary, weather forecasts, etc. For example, flipping a coin is a non-quantum mechanics macro phenomenon. There are too many factors that may determine the result, such as flipping angle, height, wind direction, and surface hardness. We cannot explain the internal mechanism of the flipping result; therefore, we cannot explain the event "the result of flipping is the tails side up."

Another critique is that the requirement of the DNP model is too high. It may make some simple explanations very complicated. For example, ice will gradually melt in the warm water, and this can be easily explained by thermodynamics. But according to Railton's DNP model, we must calculate the detailed procedure of ice melting by methods of quantum mechanics, which makes a simple explanation very complicated.

Railton's response to these criticisms is that a genuine scientific explanation is very difficult to achieve. If we cannot explain the internal mechanism, then we should not pretend to have a complete explanation. Therefore, unless we can explain the internal mechanism of the result of flipping a coin, we should not pretend to have a successful explanation. Similarly, for Railton, the event of ice melting should also be explained finally by quantum mechanics.

Section 5   Summary

Scientific explanation is a relatively new issue in the philosophy of science and is still under discussion. Professor Huaxia Zhang at Sun Yat-sen University points out that there are mainly three approaches to scientific explanation in philosophical circles. (1) *The epistemological approach* follows the epistemology developed by Hempel and makes further modifications. For example, van Fraanssen's pragmatics of explanation and Kitcher's unificationist model fall under this approach. (2) *The model approach* was proposed by the two philosophers Mary Hesse and Nancy Cartwright. They believe that we must construct models and use metaphors or analogies to understand and explain the world. (3) *The ontological approach* claims that a scientific explanation must reveal the causality and internal mechanism of a given phenomenon, and explicate its place in the whole picture and hierarchical structure of nature. Salmon and Railton's accounts both belong to this approach (Zhang, 2002, p. 32).

As the author sees it, the most important issue of scientific explanation is how to understand scientific laws. Hempel's scientific explanation models are also called covering law models, and their main feature is based on the claim that all scientific explanations – no matter whether they use the DN, IS, or DS models – must at least cover a scientific law. But what is a scientific law? Hempel attempted to propose a general form of lawlike sentences, but he finally failed to do so. Nelson Goodman also believes that there is no specific logical form of scientific laws to be found simply by analyzing counterfactual conditionals. Hence, the central concept of scientific explanation, scientific law, is ambiguous!

As a result, many problems with scientific explanation models may be due to the very nature of scientific laws. For instance, the problem of asymmetry probably arises because scientific laws are not just mathematical formulas, but also include physical interpretations. Therefore, a scientific law itself may be asymmetric. The asymmetry of laws determines the asymmetry of explanations and the structural non-identity of explanation and prediction.

In addition, the apparent need to modify the requirement of maximal specificity (RMS) is demonstrated by that fact that we usually regard the statement "salt dissolving in water is highly probable in 5 minutes" as a scientific law, yet we would never think that "hexed salt dissolving in water is highly probable in 5 minutes" or "salt and baking soda dissolving in water is highly probable in 5 minutes" are scientific laws. If we were able to clarify the nature of scientific laws, then we would be able to avoid the problem of RMS modification.

In summary, the understanding of scientific explanation involves the nature of scientific laws. To understand scientific explanations, we must first work out the nature of scientific laws. Scientific laws are not just mathematical formulas; they may also include physical interpretations. The forms of scientific laws may be various; for instance, the form of biological laws can be different from the form of physical laws. So then, are scientific laws regulative or necessary? Questions like this demand further research and analysis. Therefore, the author believes that the establishment of scientific explanation models ultimately depends on our understanding of scientific laws.

3   The very nature of laws of nature

Laws of nature have a central role in scientific research. The aim of science is to discover laws of nature. Scientists confirm or disconfirm laws through experimentation; people make use of them to improve the world. Yet, as ubiquitous as these laws are, we still must ask, what is the very nature of laws of nature? When we ask this, we find, perhaps to the surprise of some, there actually is no consensus or satisfactory answer yet. To approach this critical issue, this chapter will review the relevant literature within the philosophy of science.

Let us compare the following three statements and judge which of them are laws of nature:

(1)   The sum of the interior angles of a triangle is equal to 180°.
(2)   All metals are electrically conductive.
(3)   All money in my pocket is made of paper.

These three statements above are all universal truths, having the form "all *Fs* are *Gs*," which can be rewritten in the logical form: $\forall x(F(x) \rightarrow G(x))$. But according to our intuitive judgment, only (2) should be a law of nature. In fact, (1) is a mathematical theorem in Euclidean geometry, and (3) is only accidentally true, or, in the terminology of the philosophy of science, it is an "accidental generalization." Thus, (1) and (3) are not laws of nature.

In discussions about laws of nature in the philosophy of science, the distinction between mathematical theorems and laws of nature is an easy issue, but the question of how we distinguish laws of nature from accidental generalizations is extremely difficult to resolve. Hans Reichenbach (1947) gives two famous examples:

(4)   All solid spheres of enriched uranium have a diameter of less than 1 mile.
(5)   All solid spheres of gold have a diameter of less than 1 mile. (p. 368)

Physics tells us that uranium-235 has a critical mass. If the mass of uranium exceeds the critical mass, a chain reaction of atomic fission will happen, and that is the principle behind the atomic bomb. Thus (4) is a universal truth. As for (5), considering the fact that gold is such rare heavy metal, it is probable that the diameter of the accumulation of all gold in the universe will not exceed 1 mile. (5) is also likely to be a universal truth. However, our intuition tells us that (4) is a real law of nature, whereas (5) is just an accidental generalization. Since there is no formal difference between statements (4) and (5) except the single words *uranium* and *gold*, how can we judge that the former is a law of nature, whereas the latter is not?

Section 1   Hume's definitions of cause

David Hume discussed the nature of laws of nature, especially those associated with causality, as it was a contentious issue even in his time. For instance, his constant conjunction theory is familiar to us. However, according to John Earman (2002), an emeritus professor at the University of Pittsburgh, Hume actually proposes three definitions of causes:

> Felt determination: An object precedent and contiguous to another, and so united with it in the imagination, that the idea of the one determines the mind to form the idea of the other, and the impression of the one to form a more lively idea of the other.
>   Constant conjunction: an object precedent and contiguous to another, and where all the objects resembling the former are plac'd in a like relation of priority and contiguity to those objects, that resemble the latter.
>   Counterfactual: an object followed by another. . . where, if the first object had not been, the second never had existed.
>
> (pp. 116–117)

If Earman's interpretation is right, there can be three ways to understand Hume's view of causality: mental habit, constant conjunction, and necessity. We usually reckon Hume's view of causality to be in line with the regularity approach, especially his constant conjunction theory. However, Earman thinks those three different understandings all affect the later three approaches to laws of nature. The felt determination and constant conjunction theories belong to the regularity approach, and the counterfactual is a necessitarian approach.

Section 2   The regularity approach

Hume's most well-known view of causality is *constant conjunction theory* or *regularity theory*, which states that the reason we think there is a causal relationship between *A* and *B* is because *A* is always followed by *B*.

In *A Treatise of Human Nature*, Hume denies that the causal relationship between *A* and *B* necessitates the conclusion that if *A* happens then *B* must happen. He thinks if such "objective necessity" exists, it is either logical or non-logical. Suppose that it is logical (as mathematical theorems have logical necessity) that we cannot imagine its counter-example, such as a square circle, yet for laws of nature, it is logically possible for us to construct counter-examples. For example, objects obey the physical law of conservation of momentum during the process of collision, but we of course can imagine a situation in which the total amount of momentum increases or decreases. Thus, the law of conservation of momentum does not seem to be logically necessary. In addition, discovery of laws of nature requires empirical observations, and they cannot be *a priori* known like laws of mathematics and linguistic conventions. If objective necessity is non-logical, according to Hume's opinion, we should perceive it in our sensory experience. However, Hume denies that people have the impression of necessity in their sensory experience. Furthermore, he thinks it does not exist in our actions either, for one's will is not necessarily accompanied by his or her action. Therefore, Hume points out that the impression of necessity is simply subjective and is a fiction of the imagination.

Many scholars have inherited Hume's "felt determination," that is, the first definition of cause. For example, Nelson Goodman (1955), a professor at Harvard University, points out: "I want only to emphasize the Humean idea that rather than a sentence being used for prediction because it is a law, it is called a law because it is used for prediction" (p. 26). A. J. Ayer (1956), a professor at Oxford University, suggests "that the difference between our two types of generalization lies not so much on the side of the facts which make them true or false, as in the attitude of those who put them forward" (p. 162). Professor Nicholas Rescher (1970) at the University of Pittsburgh also writes in a similar vein: "Lawfulness is not found in or extracted from the evidence, it is superadded to it. Lawfulness is a matter of imputation" (p. 107).

According to these advocates, the reason why "the diameter of a solid sphere of enriched uranium is less than 1 mile" can be a law is because of our attitude toward it: using the generalizations for prediction or imputation. While the reason why "the diameter of a solid sphere of gold" cannot be a law is also because of our attitude toward it: not using the generalizations for prediction or imputation.

However, this approach is thought to be so subjective that it is not widely accepted. Another school prevailed and has finally become the representative of the regularity approach. This school is the so-called systematic approach, and its advocates think that laws of nature should be part of a "best deductive system."

For example, J. S. Mill (1904) proposes two expressions to clarify the question "what are laws of nature?" One possible expression, as Mill sees it, can be: "What are the fewest and simplest assumptions, which being granted, the whole existing order of nature would result?" Another is: "What are the fewest general propositions from which all the uniformities which exist in the universe might be deductively inferred?" (p. 230).

Frank Ramsey (1978) thinks, in a more specific way, that "Laws were consequences of those propositions which we should take as axioms if we knew everything and organized it as simply as possible in a deductive system" (p. 138).

Again, recall the examples of the diameter of the solid spheres of uranium and gold. It is possible for us to deduce the critical mass of the solid spheres of uranium from the principles of quantum mechanics. Thus, the statement "All solid spheres of enriched uranium have a diameter of less than 1 mile" is part of our current best deductive system, and, in this way, is a law of nature. However, from our current knowledge, we cannot deduce the critical mass of gold, so the statement "All solid spheres of gold have a diameter of less than one mile" is not part of our best deductive system, and it is simply an accidental generalization.

Section 3   Problems of the regularity approach

F. Dretske (1977), in interpreting the regularity theory, states, "law is the compound of universal truth and some special function or status, X, which is the qualification for universal truth to be a law," that is, "law = universal truth + X" (p. 251). Regularity theorists provide various answers for this X, and Dretske summarizes them as the following:

(1) High degree of confirmation
(2) Wide acceptance (well established in the relevant community)
(3) Explanatory potential (can be used to explain its instances)
(4) Deductive integration (within a larger system of statements)
(5) Predictive use (p. 251)

Dretske (1977) also points out that the six problems of the regularity approach are as follows:

(1) A statement of law has its descriptive terms occurring in opaque positions.
(2) The existence of laws does not await our identification of them as laws. In this sense, they are objective and independent of epistemic considerations.
(3) Laws can be confirmed by their instances, and the confirmation of a law raises the probability that the unexamined instances will resemble

(in the respect described by the law) the examined instances. In this respect, they are useful tools for prediction.

(4) Laws are not merely summaries of their instances; typically, they figure in the explanation of the phenomena falling within their scope.

(5) Laws (in some sense) "support" counterfactuals; to know a law is to know what would happen if certain conditions were realized.

(6) Laws tell us what (in some sense) must happen, not merely what has and will happen (given certain initial conditions). (p. 260)

Universal truths have the property of transparency, and transparency implies that properties have transitivity. For example, we can infer $\forall x(Hx \to Gx)$ from two premises, $\forall x(Fx \to Gx)$ and $\forall x (Fx \leftrightarrow Hx)$. In a special case, we can infer that "The largest mammal lives in the sea" is a universal truth from two universal truths "balaenoptera musculus lives in the sea" and "balaenoptera musculus is the largest mammal." However, laws of nature are opaque, which is opposite to the property of transparency. For example, the two statements "balaenoptera musculus lives in the sea" and "balaenoptera musculus is the largest mammal" are both laws of nature, while "the largest mammal lives in the sea" is not. Therefore, Dretske draws a conclusion that the regularity approach cannot work out the problem of opacity. We can only deal with it if we introduce the modal concept of opacity.

Assume we have such a law of nature as "If we toss a coin, we will get the side of the national emblem." When we actually toss a coin 9 times and get the side of the national emblem 9 times, these 9 tosses are confirmation of this law. We may infer that the next time, the possibility that we will get the national emblem side of the coin will increase. If it is not a law of nature, those 9 tosses are no longer confirmation, and the possibility of the next toss resulting in the national emblem side up may remain 50%. Thus, Dretske also thinks that the regularity approach cannot resolve the problem of confirmation and prediction of laws.

Besides, in Dretske's view, Ayer confuses the difference between epistemological problems (Why do we believe a law is a law?) and ontological problems (What is a natural law?). A law of nature can remain undiscovered for a long time, but it always exists independent of our knowledge and language. However, Ayer's *epistemological regularity theory* requires that only when scientists treat a universal truth as a natural law does it become a natural law. This logically implies there are no unknown natural laws, although Ayer denies this implication. This, however, seems inconsistent with our intuitions, for, before Newton's discovery, people did not know the law of universal gravitation. So, is he claiming that this law was nonexistent before Newton?

Section 4   The necessitarian approach

Dretske emphasizes the necessity in laws of nature. Thus, he, along with D. M. Armstrong and M. Tooley, takes the *necessitarian approach* to laws of nature. More specifically, Dretske (1977) proposes the "universals theory," claiming that "statements of natural laws are not the universal generalizations of particulars, but the singular statements of universals." He continues:

> Law-like statements are singular statements of fact describing a relationship between properties or magnitudes. . . . Laws are (expressed by) singular statements describing the relationships that exist between universal qualities and quantities; they are not universal statements about the particular objects and situations that exemplify these qualities and quantities.
>
> (pp. 253–254)

Dretske rewrites the form of natural laws "All *Fs* are *Gs*" as

F-ness $\rightarrow$ G-ness

Both *F*-ness and *G*-ness are universals. For example, all metals have the universal property of "metallicity," and all conductors have the universal property of "electrical conductivity." In this equation, "$\rightarrow$" is not a material implication, as in propositional logic, but is to be understood as "yields" or "generates." Thus, "*F*-ness $\rightarrow$ *G*-ness" is read as "The property of being *F* necessitates the property of being *G*." For example, "All metals are electrically conductive" should be formulated as "Metallicity necessitates electrical conductivity." Dretske thinks that since *F*-ness necessitates *G*-ness, then if *x* is *F*, *x* must be *G*. For example, if *x* has the property of metallicity, *x* must have the property of electrical conductivity.

Although Dretske cannot provide a proof for this necessity, he presents an analogy to explain his view. The US Constitution prescribes that the president, as the head of executive authority, must act in conformity with the acts passed by the Congress as the legislative authority. The law prescribes this, so no matter who is president, he or she must be accountable to Congressional actions. Abraham Lincoln, John Kennedy, Ronald Regan, and William Clinton all behaved in this way. Even if the dictator of Iraq, Saddam Hussein, had been elected the president of the United States, he would have also been accountable to the acts of Congress.

In Dretske's view, in natural laws, *F*-ness and *G*-ness are like offices, and no matter what has the property of *F*-ness (or whomever enters *F* office), it must have the property of *G*-ness (be responsible for *G* office). This analogy demonstrates the place of necessity in laws of nature.

Section 5　Problems of the necessitarian approach

Dretske's *universals theory* presupposes the existence of universals like "redness," "electrical conductivity," "metallicity," etc. He thinks that "universal properties exist, and there exists a definite relationship between these universal properties, if there are any laws of nature" (Dretske, 1977, p. 267). However, many empiricist philosophers have questioned these invisible and untouchable "universals." Van Fraassen further points out that there are two difficulties in the necessitarian approach: the *identification problem* and the *inference problem*.

First, how do we recognize the necessity, namely the nomic necessity, in the necessitarian approach? Actually, Hume has already given this a clear analysis: either nomic necessity is logical necessity, but scientists do not think natural laws are logical truths; or nomic necessity is not logical necessity, and then it is totally subjective and we are left with the question of how we can justify the objective necessity.

Second, the inference problem stems from the fact that premises "F-ness $\rightarrow$ G-ness, and a is F" can only infer that "a is G," but not "a necessarily is G," thus showing that Dretske's inference is not valid. Van Fraassen also thinks that the office analogy is not incorrect. The fact that US presidents are constrained by the Congress is provided by the US Constitution, so they must execute the acts passed by the legislature. However, for the "F" and "G" in a law of nature, is there anything fundamental to constrain the laws? In light of this, van Fraassen thinks Dretske's necessitarian approach is impracticable.[1]

Besides, D. H. Mellor also criticizes the necessitarian approach. He follows Ramsey's opinion, holding the view that "particulars" (special bodies, occupying space-time) and "universals" (general properties, not occupying space-time) are not two totally different kinds of things. According to Wittgenstein's claim in the *Tractatus* that "The world is the totality of facts, not things," Ramsey thinks that particulars and universals should not be seen as different dependent properties, but necessary parts of facts, which means particulars and universals are simply different aspects of facts. Thus, there must be an instance of "F-ness $\rightarrow$ G-ness." However, laws of nature allow no such instance. For example, Newton's first law states that "when viewed in an inertial reference frame, an object either remains at rest or continues to move at a constant velocity, unless acted upon by a force" (Browne, 1999, p. 58). However, in nature, there is no such an object that is either not acted upon by some force or in the state of force balance.

Mellor also argues against S. Kripke and H. Putnam's "metaphysically necessary" account of laws of nature. He criticizes them for using modal logics and the concept of "possible worlds" to explain "necessity" in laws of nature, which commits the logical fallacy of begging the question (Mellor, 1998, pp. 846–864).

Section 6   Other possible approaches

Van Fraassen thinks that symmetry should replace the central role of laws in science and philosophy. In *Laws and Symmetry*, he has three main objectives: first, to show the failure of current philosophical accounts of laws of nature; second, to refute arguments for the reality of laws of nature; and third, to contribute to an epistemology and a philosophy of science antithetical to such metaphysical notions (van Fraassen, 1989).

He claims that what modern physics discusses is symmetry and continuity, not universality or necessity, the natural kind or the very nature, the contingent or accidental. *Laws of nature* is a concept of degeneration in contemporary science. *Symmetry* refers to the fact that after translation and specular reflection, a physical phenomenon remains the same. However, physical phenomena cannot satisfy all the symmetries. For example, the motions of two atoms as mirror images of each other caused by a weak interaction does not satisfy the mirror symmetry. A famous concrete example is the decay of Co-60, which is the experimental confirmation of the nonconservation of parity discovered by Chen-Ning Yang and Tsung-Dao Lee. But physical phenomena should satisfy most kinds of symmetries, meaning that symmetry is a necessary condition. *Continuity* implies that there is no action at a distance, so causality should be continuous in space-time.

Ronald Giere received his PhD from Cornell University in 1968 and has been teaching at the University of Minnesota. He used to be the director of the Minnesota Center for Philosophy of Science, and now is a Professor Emeritus. His monographs include *Understanding Scientific Reasoning* (1979), *Explaining Science* (1988), *Science without Laws* (1999), and *Scientific Perspectivism* (2006). Important to this discussion, Ronald Giere also argues against laws of nature, suggesting in *Science without Laws* that there exists regularity and necessity in nature, but no laws.

Giere summarizes the concept of laws in modern science from a historical perspective: Galileo believed God was the author of *the Bible* and "book of nature," but that in *Discourses and Mathematical Demonstrations Relating to Two New Sciences*, he seldom mentioned the role of laws. Robert Boyle shared many of Newton's theological opinions; however, he was cautious to use the idea of laws because, he thought, only beings with consciousness know how to use laws. Johannes Kepler believed in both God and laws of nature. Descartes and Newton firmly believed that laws of nature were stipulations from God for all actions in nature because God is the creator of the universe. Thus, laws of nature are universally true, necessary in an absolutely obligatory sense, and independent from human beliefs. Humans not only obey God's laws of nature, but they also obey God's moral laws. In his later years, Newton even attributed the universality of laws of motion to God. Thus, in early times, the concept of "law of nature" was closely connected with God. Not until the Law of Natural Selection was proposed by Darwin in *The Origin of Species* did laws of nature begin to be separated from God's will (Giere, 1999, pp. 84–96). Giere's summary shows, from the origin, that the concept of laws of nature are closely affected by theology. Therefore, it is reasonable for us to give up this concept.

Without laws, however, what will science be like? Giere advocates the layered structure of an "equation–model–world." For example, the basic foundation of quantum mechanics is Schrödinger's equation, and based on this equation, we can construct the model of metal's electron sea. Furthermore, we can explain the phenomenon of electrical conductivity of metals. Thus, we get the following schema:

Schrödinger's equation (equation)

↓

Model of metal's electron sea (model)

↓

All metals are electrically conductive (real world).

From the view of the history of science, Giere shows in *Mathematical Principles of Natural Philosophy* that principles are what people use to model and then represent rules of special cases in nature. Thus, the layered structure of science can be also presented by "principle–rule–modeling."

Sandra Mitchell, the chair of the Department of History and Philosophy of Science at the University of Pittsburg, denies the simple dichotomy of natural laws and accidental generalizations but emphasizes the dimensions of laws of nature. She proposes three approaches to natural laws: (1) the normative approach, which stipulates the normative features of laws of nature, then checks those generalizations in specific sciences according to the definition; (2) the paradigmatic approach, which enumerates the paradigmatic laws in science, and then finds the common features between them; (3) the pragmatic approach, which first studies the pragmatic effects of laws in scientific research, and then examines whether or not these generalizations play the same role in special sciences (Mitchell, 1997, p. 469).

If we follow the normative approach, scientific laws are usually thought to have four main features: (1) logical contingency (have empirical content); (2) universality (cover all space and time); (3) truth (exceptionless); (4) natural necessity (not accidental) (Mitchell, 2000, p. 246). According to this definition, there might not be any laws in the special sciences, such as biology. Mitchell prefers the paradigmatic approach, and especially the pragmatic approach, proposing that laws can vary to different degrees. For example, there exists continuity from paradigmatic ideal laws (such as the law of mass-energy conservation) to paradigmatic accidental generalizations (such as, the coins in my pocket are all made of copper):

> Law of mass-energy conservation
> Law of conservation of mass
> 2nd Law of Thermodynamics
> Periodic law
> No uranium-235 sphere has diameter greater than 100 meters.
> Galileo's law of free fall
> No gold sphere has diameter greater than 100 meters.
> Mendel's law of independent assortment
> All the coins in Goodman's pocket are made of copper.
>
> (Mitchell, 2000, p. 253)

Mitchell (2000) proposes a multi-dimensional conceptual framework of scientific laws, for example, stability and strength in the dimension of ontology, degree of abstraction, simplicity, cognitive manageability, etc. (pp. 262–263). She uses 3-dimensional space to denote the 3 dimensions in scientific laws: stability, strength, and degree of abstraction. All lawlike statements, from "conservation of mass-energy" to the "copper coins," can be represented in that 3-dimensional space. The statement of conservation of mass-energy is the highest in all three degrees, while the statement of copper coins is much lower.

Section 7   Summary

From the perspective of the regularity approach, laws of nature are the real descriptions of how objects actually act. However, the necessitarian approach assumes that laws of nature not only describe what the world is, but that they also claim what the world must be. These two approaches both have their own advantages, as well as their own disadvantages that are hard to overcome. So, what about other possible approaches?

Van Fraassen and Giere both believe there are no laws of nature. However, if, as they assume, the goal of science is to know the world and promote the world, they must appeal to concepts other than laws. Van Fraassen introduces symmetry, which is a necessary condition for physical laws but is not sufficient. Physical laws must satisfy some symmetries, but not all phenomena in physics, which satisfy symmetries, can be admitted as laws. Therefore, van Fraassen's concept of symmetry may not be sufficient to uncover the very nature of laws of nature.

Giere turns to equations (or principles) and models. But these new concepts are just other names of laws of nature, not something essentially new. Our original question is whether the nature of laws of nature is either one of regularity or necessity. If there are simply equations and principles in science, can we still ask whether the nature of equations and principles is one of either regularity or necessity? Thus, Giere does not dispel the question, but rather changes the question into a new form.

It seems that Mitchell's approach is more plausible. However, she denies the dichotomy of genuine laws of nature and accidental generalizations. She refuses the necessity in laws of nature, yet this is something the author affirms. For example, it is impossible for us to invent a perpetual-motion machine, the velocity of any object is never more than the velocity of light, the dead cannot come back to life, etc. Thus, Mitchell may meet difficulty when arranging the laws of nature and accidental generalizations in her multi-dimensional conceptual framework. She may think the statement "All solid spheres of gold have a diameter of less than 1 mile" holds throughout the whole universe, while Mendel's law holds only in special species on the Earth. Thus, she prefers to believe the former statement is more lawlike than the latter. However, most scientists prefer that Mendel's law be considered the natural law rather than the statement "All solid spheres of gold have a diameter of less than 1 mile," for Mendel's law expresses biological necessity to some degree.[2]

The author thinks that the failure of Mitchell's dichotomy requires us to come back to Hume. The necessity of laws of nature is stipulated by nature, which is an ontological problem, and the necessitarian approach provides a good solution. Laws of nature are the most coherent system we use to explain and promote the world. Moreover, the regularity approach, to its credit, provides a good answer for the epistemological approach to laws of nature.

If, in epistemology, we can only hold the regularity approach to laws of nature, how can we know the necessity of laws in ontology? Perhaps the idea in Chinese philosophy of "expanding one's personality and knowing the degrees of heaven" can provide a solution. Laws of nature are always formulated as $\forall x(Fx \rightarrow Gx)$, and this formula is logically equivalent to $\neg\exists x(Fx \wedge \neg Gx)$, but how can we know all Fs must be Gs? A possible way to deal with this is to find the things that "are of F but not of G." For example, people have tried thousands of times to invent a perpetual-motion machine but have always fail, so we know that the laws of conservation and transformation of energy are necessary stipulations in nature. In Chinese history, many emperors who strived to find the elixir of immortality all failed. Then people realized the fact that we cannot come back to life once we are dead is a biological law. Of course, this way of discovering the necessity of natural laws is difficult. Nonetheless, Confucius, the ancient Chinese sage, stated, "At 15, I had my mind bent on learning"; however, "At 50, I knew the degrees of heaven." As for those of us normal people who are not sages, we still have a long way to go in our attempt to deepen our insights into the laws of nature.

Notes

1 Van Fraassen's PhD supervisor Adolf Grünbaum said, in a conversation with me, Dretske's analogy was not correct. Historically, many US presidents actually did not obey the Congress, instead finding ways to control it.
2 The author asked Mitchell about this question in a private conservation, and she replied that she was in a hurry writing that paper, so she made the mistake when arranging the order. This seems to reveal that she tried to cancel the dichotomy between laws of nature and accidental generalizations, which, as a result, creates conflict.

4 The conceptions of scientific explanation and approaches to laws of nature

Section 1   Background

Carl G. Hempel proposes three models of scientific explanation, namely, the DN model, the IS model, and the DS model. These models must include certain scientific laws; thus they are usually referred to as covering law models. However, these models have encountered a series of difficulties, for instance, the problem of asymmetry, the irrelevance problem, the problem of requirement of maximum specificity, and the problem of epistemic relativity of statistical explanation. As a result, the covering law models have been questioned more and more.

Wesley Salmon summarizes the developmental trends of scientific explanation into the following three conceptions: (1) the epistemic conception can be stated as "explaining an event is equal to showing its nomic expectability," following (revising and consummating) the epistemic approach pioneered by C. G. Hempel. Examples of this conception are van Fraassen's pragmatics of explanation and Philip Kitcher's unificationist models. (2) The modal conception can be stated as "scientific explanations aim at showing that the explained events must happen." For example, D. H. Mellor and others have proposed a modal interpretation for probabilistic explanations. (3) The ontic conception is stated as "scientific explanations aim at revealing causality and the intrinsic mechanism of the explained phenomena, and clarifying the status of it in the overall natural picture and hierarchical structure." Two examples of the ontic conception are W. Salmon's causal theory and P. Railton's DNP model (Salmon, 1989, pp. 120–122).

In addition, there are some other popular views of scientific explanation. (4) As for deductivism, John Watkins proposes that all explanations are deductive inference. (5) Finally, for the simulacrum account, Nancy Cartwright insists that the route from theory to reality is from theory to model, and then from model to phenomenological law; on this view, the phenomenological law is indeed true of the objects in reality – or at least could be – but the fundamental laws are true only of objects in the model.

In the third chapter of this monograph, we mentioned that there are two main approaches to laws of nature: (1) J. S. Mill, Frank Ramsey, and David Lewis' regularity approach insists that the laws of nature only show regularity and that a contingent generalization is a law of nature if and only if it appears as a theorem (or axiom) in each of the true deductive systems that achieves a best combination of simplicity and strength. (2) D. M. Armstrong, F. Dretske, and M. Tooley's necessitarian approach insists that the laws of nature show necessity and are usually expressed by singular statements describing the relationships that exist between universal qualities and quantities.

This chapter will attempt to show that there is a close connection between the conceptions of scientific explanation and approaches to laws of nature, as well as how philosophers' understanding of laws of nature greatly influences their conceptions of scientific explanation. The author will try to expound upon these connections in detail and investigate whether or not they are logically consistent. Finally, some opinions and rebuttals will be offered.

Section 2   The pragmatics of explanation and the no-laws argument

Van Fraassen's pragmatics of explanation argues that explanation is the relation between theories, facts, and contexts. In his point of view, scientific explanation is to an answer a why-question. These kinds of questions can be formulated as the form $Q = <P_k, X, R>$, in which P_k is the topic of the question, $X = \{P_1, P_2, \ldots, P_k, \ldots\}$ is the contrast-class, and R is the relevance relation. Scientific explanation tries to answer such questions: A bear R with respect to $<P_k, X>$. Thus, there is no particular reason that scientific explanation should not appeal to laws of nature.

There is a close connection between van Fraassen's pragmatics of explanation and his attitude toward laws of nature. In *Laws and Symmetry* (1989), van Fraassen claims, "Modern physics focuses on symmetry and continuity, rather than universality or necessity, natural classes or essence, accident or accidental generalization. Laws of nature as a concept is degenerative in contemporary sciences." He proposes the use of the concept of symmetry as a replacement for the concept of laws of nature. Since there are no longer scientific laws under this conception, what is left in scientific explanation is pragmatics, and we need not investigate the relation between explanations and laws.

However, P. Railton points out that we must distinguish ideal explanatory text from explanatory information (Railton, 1981). Using Hempel's example of a frozen crack in a car radiator, the ideal explanatory text should include general laws such as water freezing and expanding and initial conditions, such as the fact that the car radiator was full of water, the outdoor temperature was lower than zero degree centigrade, etc. Finally, we can derive from this that there was a frozen crack in the car radiator. Additionally, the explanatory information depends on the context, as one may be curious as to why Hempel's, rather than his neighbor's, car radiator cracked, while others will want to know why the car's radiator cracked, rather than the car's fuel tank. We should answer these questions respectively; this is the pragmatics of explanation. There is no contradiction between the logic and the pragmatics of explanation. Instead, they are complementary to each other.

W. Salmon also presents that criticism that van Fraassen's concept of the relevance relation R is ambiguous. He writes, "if R is not the real correlativity, then the correlation of A and P_k is just a sarcastic remark" (Salmon, 1989).

Van Fraassen insists that there is no law, so there is no logic of explanation, but in the author's opinion, the contraposition of his proposition is that if there is a logic of explanation, there must be laws. In fact, in a visit to the University of Pittsburgh in 2005–2006 academic year, the author asked van Fraassen's PhD supervisor Adolf Grünbaum about this issue. He agreed that van Fraassen denies the concept of laws of nature, but if we study such scientific concepts as explanation, we must seek a substitute for laws of nature.

Section 3   The unificationist approach and the MRL view

In the epistemic conception of scientific explanation, P. Kitcher's unificationist approach is worth mentioning. This approach was originally raised by Michael Friedman in 1974 and was then developed by P. Kitcher in 1989. Kitcher argues that science advances our understanding of nature by showing us how to derive descriptions of many phenomena, using the same pattern of derivation again and again. In demonstrating this, it teaches us how to reduce the number of types of fact that we accept as ultimate (Kitcher, 1989).

Kitcher's famous slogan is "only connect." He also proposes the concept of the explanatory store $E(K)$. For every phenomenon K to be explained, if $E(K)$ can do best unify K, then $E(K)$ constitutes an explanatory store for K. Because of this, W. Salmon calls P. Kitcher's approach top down.

Kitcher's explanatory schema can be expressed as <schematic argument, filling instructions, classification>. The schematic argument includes sentences in which some non-logical expressions are replaced by dummy letters, which can carry several values. The filling instructions are directions for replacing the dummy letters of the schematic sentences with their appropriate values. The classification is a set of statements that describes the inferential structure of the schema. The first and second parts of the schema are the premises, and the third is the conclusion.

Kitcher uses atomic chemistry as an example:

(1) The compound Z between X and Y has an atomic formula: X_pY_q.
(2) The atomic weight of X is x and the atomic weight of Y is y.
(3) The weight ratio of X to Y is $p_x : q_y$ ($= m : n$).

The first two sentences are the premises, and the last one is the conclusion. This scheme can be used to analyze atomic weight ratio. Similarly, we can use Newtonian mechanics to unify various phenomena such as the planetary orbits of the solar system, tides, free-falling bodies, pendulums, etc. Thus, we acknowledge that Newtonian mechanics successfully explains these natural phenomena.

If we try to find the relation between the unificationist account of scientific explanation and the MRL view of laws of nature, it should be a perfect match, as both of them emphasize the importance of fundamental laws in the unification or explanation of various kinds of phenomena. In the unificationist account, the fundamental laws are the best unifiers to save various phenomena, while in the MRL view, the fundamental laws are the axioms or most important theorems of our knowledge system. Therefore, various kinds of events or regularities may be derived from the laws.

Consequently, according to Stathis Psillos (2002):

> Both of these are brought together under a central concept: unification. As regards the nature of explanatory dependence, it is still derivation; but, not any kind of derivation. It is derivation within a maximally unified theoretical system. As regards the nature of the best deductive systematization, it too is the systematization that maximally unifies a theoretical system. Explanation and best deductive systematization become one.
>
> (p. 264)

The basic unifiers are the most fundamental laws of nature!

Section 4   Modal conception and the necessitarian approach

According to Salmon, D. H. Mellor gives a modal interpretation of probabilistic explanation. Mellor thinks there are degrees of necessitation. If an event is completely necessitated, its occurrence is fully entailed by laws and explanatory facts. If it is not fully necessitated, its occurrence is partially entailed by explanatory facts. The greater the degree of partial entailment, the better the explanation. Mellor finally agrees that within scientific explanation, events to be explained are completely or partly restricted by laws and explanatory facts (Mellor, 1976).

There may also be a match between the modal conception of scientific explanation and the ADT view of laws of nature, since both of them appeal to nomic necessity. Surprisingly, concerning the laws of nature, Mellor follows Ramsey's view and argues against the necessitarian approach. He appeals to Carnap's inductive logic to justify his conception, on appeal with which Salmon disagrees. Mellor writes,

> I follow Ramsey in taking particulars and universals to be simply parts of facts picked out in order to generalize. . . . If the law is vacuous, there are no such facts; and no facts leave no residue. If there are no *Fs*, there is no *F*. So there will be no fact *FNG* to make such a vacuous law true, and the Armstrong-Dretske-Tooley theory fails.
>
> (Mellor, 1998, p. 862)

Mellor prefers a modal conception of scientific explanation but argues against the necessitarian approach to laws of nature. This does not appear to be coherent when making cross-field comparisons between scientific explanation and laws of nature. However, Mellor would still agree with the covering law thesis because he insists that an event-to-be-explained is (fully or partially) entailed by laws and explanatory facts (Mellor, 1976). Thus, he thinks explanation must appeal to laws, though his conception of explanation seems to be against his understanding of laws of nature.

Section 5   The ontic conception, causality, and laws

The two most important representatives of the ontic conception are W. Salmon and Peter Railton. Salmon is famous for his comprehensive review and critical comments of scientific explanation. He first proposed the statistical relevance model (SR model) and later developed the causal mechanism model (CM model). Salmon states,

> Explanatory knowledge is knowledge of the underlying mechanism – causal or otherwise – that produce the phenomena we want to explain. To explain is to expose the internal workings, to lay bare the hidden mechanisms, to open the black boxes nature present to us.
>
> (Salmon, 1988, p. 160)

This is why his slogan is "put cause back into because."

Thus, Salmon advocates that statistical relevance, rather than the requirement of high probability, is the core of statistical explanation. Statistical relevance can be formulated as: prob(E/C) > prob(E/non-C), and there are no more factors H by which E is blocked from C, namely, prob($E/C\&H$) = prob(E/H).

Salmon emphasizes the relation between explanation and causation: scientific explanations only show how events fit into the causal structure of the world. While early Salmon distinguished causal processes from pseudo processes in terms of the ability to *transmit marks*, later Salmon defined a causal process as a process that transmits a non-zero amount of a *conserved quantity* at each moment in its history (Woodward, 2009). Salmon also calls his scientific explanation a bottom-up approach, which takes causal relations to be prior to relations of explanatory dependence.

What is causation, then? Salmon suggests using the concept of causal processes to replace the traditional notion of causal events. Some causal processes are real (for example, an earthquake causes a tsunami), while some are pseudo (for example, the movement of a light cross on a cloud caused by a searchlight). Real causal processes cannot travel faster than the speed of light; otherwise, it would contradict Einstein's theory of relativity. So the tsunami always comes after the earthquake, but not vice versa. Pseudo processes can exceed the velocity of light, however. For example, if a searchlight is powerful enough and moves fast enough, while a cloud is far enough away, we can achieve super light speed movement of the light cross.

W. Salmon distinguishes causal processes from pseudo processes as follows: causal processes are capable of transmitting marks, while the latter cannot. As the author sees it, though, Salmon's use of mark transmitting as a criterion to distinguish causal processes may pose the problem of begging the question. For detailed arguments, please see Chapter 5.

Another representative of the ontic conception, Peter Railton (1978), proposes the Deductive-Nomological Model of Probabilistic Explanation (DNP). Here is the schema of the DNP:

(A) A theoretical derivation of a probabilistic law of the form (b)

(B) $\forall x \forall t [Fx,t \rightarrow P(Gx,t) = r]$ probabilistic law
(C) Fe, t_0 initial condition
(D) $P(Ge, t_0) = r$ statement of a single-case propensity

(E) Ge, t_0 parenthetic addendum

Railton's point is often called the "nomothetic account" because he finds that in Hempel's concept of "nomic expectability," nomicity and expectability can conflict with each other. For example, in low-probability explanations, we can find nomic relations between explananda and explanans, although we may not be able to infer the explanandum from the explanan. Railton's preference is nomicity, so he pays more attention to finding the nomological relations in scientific explanation, while explananda need not be highly expectable, deductive or highly inductive, from explanans. He thinks explanatory practice in the sciences is a central way of law seeking.

Railton (1981) insists, "Their [ideal explanatory text] backbone is a series of law-based deductions" (pp. 248–249). According to Psillos (2002) again, "this ideal text may well be the text that the web-of-laws [MRL] approach to laws" (p. 260). Here we find that Railton accepts the importance of laws of nature in scientific explanation. He regards scientific explanation as law-based deduction, so his understanding of laws of nature should be close to that found in the MRL view, the best deductive system of knowledge.

The ontic conception of explanation puts emphasis on the role of laws in scientific explanation. Even Salmon (1989), who favors causal models, also admits the following:

> According to the ontic conception, the events we attempt to explain occur in a world full of regularities that are causal or lawful or both. These regularities may be deterministic or irreducibly statistical. In any case, the explanation of events consists in fitting them into the patterns that exist in the objective world.
>
> (pp. 120–121)

Salmon also calls the ontic conception objectivism. Objectivism means that subjective factors shall be suspended; for example, objectivists oppose the ambiguity of statistical explanation. Salmon points out that objectivists insist on the covering law's conception of scientific explanation, but between causation and laws, he prefers causation.

In summary, both Salmon and Railton emphasize the essentiality of laws of nature, or regularities, in scientific explanation in an objectivist sense. Thus, their ontic conception should be compatible with Hempel's covering law thesis. Salmon regards causation as prior to laws of nature, so he prefers causal explanation to lawful explanation, while Railton's understanding of laws of nature is close to the MRL approach, leading him to insist on nomicity, rather than Hempel's nomic expectability, in scientific explanation.

Section 6   Deductivism and laws

John Watkins believes all scientific explanations must be deductive, therefore only the DN and DS models qualify, while the IS model does not. He
writes:

> Both explain empirical regularities by appealing to higher level struc-
> tural laws that are taken as immutable and absolute. In the case of
> classical physics, such a law says that, given only that such-and-such
> conditions are satisfied, nothing whatever can prevent a certain out-
> come from following. In the case of microphysics, it says that, given
> only that such-and-such conditions are satisfied, nothing whatever can
> alter the chance that a certain outcome will follow. One could as well
> call the latter an "iron law" of chance as the former as "iron law" of
> nomic necessity. Both kinds of law, in conjunction with appropriate ini-
> tial conditions, can explain empirical regularities and, indeed, singular
> macro-events (provided that the macro-event in question is the result-
> ant of a huge aggregate of micro-events. . .).
>
> (Watkins, 1984, p. 246)

We may find that Watkins should agree with the covering law thesis
because he insists that only iron laws – either of nomic necessity or of
chance – explain. However, there are two potential objections to Watkins'
deductivism.

First, Watkins' understanding of iron laws is very rigid. If we consider
ceteris paribus conditions or Hempel's concept of provisos (Hempel,
1988), laws of nature need not be so inflexible. In recent years, the traditional understanding of laws of nature as universal and exceptionless has
softened. Many philosophers of science have begun to accept so called *ceteris paribus* (CP) laws, though some philosophers do continue to deny them
(Reutlinger, Schurz, and Hüttemann, 2011). For example, Mendel's Law
of Segregation in biology or the Law of Supply and Demand in economics
only hold if other factors remain equal or there is no interference. These are
the so-called CP conditions. Neither of the laws is iron in Watkins' sense,
yet we can still use the CP laws to explain some biological and economic
phenomena.

Second, in some scientific explanations, explananda can be derived or calculated from explanans. However, the derivation or calculation may involve
approximation or idealization, which need not be deductive. So Watkins'
deductivism may exclude many real explanations in scientific practice.

In one word, Watkins insists on the covering (iron) law thesis of scientific
explanation. The author's only two objections are (1) his rigid understanding

of laws of nature leads to his strict deductivism, and (2) his insistence on deductive explanation is too formal for scientific practice.

Section 7   The simulacrum account and laws

Nancy Cartwright's simulacrum account is also quite influential in contemporary thought. She thinks the route from a theory to reality is from theories to models, and then from models to phenomenological laws. The phenomenological laws are indeed true of the objects in reality – or, at least, might be true – but the fundamental laws are true only of objects in the model (Cartwright, 1983). Here is the picture of her scientific explanation:

> Fundamental laws
> Models
> Phenomenological laws
> Reality

For example, Schrödinger's equation will be true in the electron-sea model, which can provide the phenomenological law "All metals are conductors." Then the phenomenological law explains why my steel watch can conduct electricity. So, the example of her explanation should be:

> Schrödinger's equation (fundamental law)
> Electron-sea model (model)
> All metals are conductors. (phenomenological law)
> The steel watch can conduct electricity. (reality)

Cartwright disagrees with Hempel's models, especially on the facticity of fundamental laws. In her first book *How the Laws of Physics Lie*, Cartwright quotes from the *Oxford English Dictionary* that a simulacrum is "something having merely the form or appearance of a certain thing, without possessing its substance or proper qualities." She uses the word *model* deliberately to suggest the failure of exact correspondence. In addition, Cartwright thinks her simulacrum account is not a formal account, which does not need to be a deductive argument.

However, Cartwright would still agree with the role of scientific laws in science. She suggests that different incompatible models are used for different purposes, and this adds, rather than detracts, from the power of the theory. In 1999, when she published *The Dappled World*, her claim became more moderate. "Nowadays I think that I was deluded bout the enemy: it is not realism but fundamentalism that we need to combat" (Cartwright, 1999, p. 23). She no longer denies the reality of laws; instead, she is against the universality of fundamental laws. She proposes a concept of a "patchwork of law" and also a "nomological machine," which is

> a fixed (enough) arrangement of components, or factors, with stable (enough) capacities that in the right sort of stable (enough) environment will, with repeated operation, give rise to the kind of regular behaviour that we represent in our scientific laws.
>
> (Cartwright, 1998, p. 2)

As the author posits, it may be that Cartwright proposes the simulacrum account because she wants to emphasize that all laws of nature are *ceteris paribus* and that the capacity or nature in a nomological machine is more basic than laws of nature. As this reveals, her conception of scientific explanation is closely related to her approach to laws of nature.

Section 8   Summary

According to Salmon, there are four agreements and three problems in the study of scientific explanation. The four agreements are the following. (1) Science can teach us not only that, but also why. (2) The "received view" of the mid-1960s is not viable. (3) What constitutes a satisfactory scientific explanation depends upon certain contingent facts about the universe. (4) The pragmatics of explanation was not accorded sufficient emphasis in the "received view." The three problems are (1) lawlike statements and purely qualitative (or projectable) predicates; (2) the problem of causality; and (3) remote correlation (i.e., the Einstein-Podolsky-Rosen paradox) (Salmon, 1989, pp. 180–186).

Salmon's first problem concerns the nature of scientific laws, while the last two are about causation. If we consider the close connection between laws and causation, we shall find the three problems are in fact are concerned with the same question: What is causal nomicity?

Following is a brief summary of the five conceptions of scientific explanation and their understanding of scientific laws (1) van Fraassen denies laws of nature in science, so he proposes the pragmatics of scientific explanation; however, it is possible there is only a verbal difference between his concept of symmetry and the traditional concept of laws of nature. If we further ask what his concept of relevant relation is, we may finally end up appealing to laws. Kitcher's unificationist account perfectly matches with the MRL view in the regularity approach to laws. (2) Mellor holds a modal interpretation of probabilistic explanation, but that is inconsistent with his regularity approach to laws. (3) Salmon argues that causal or lawful nomicity is essential in scientific explanation and Railton's view of laws belongs to the regularity approach. Their ontic conception should be compatible with Hempel's covering law thesis. (4) Watkins's deductivism is even more rigidly insistent on general laws or probability laws. (5) Cartwright argues for the informal simulacrum account because she prefers *ceteris paribus* (CP) laws to the universality of fundamental laws.

The author suggests two conclusions. (1) The above conceptions of scientific explanation are closely related to the approaches to laws of nature. The philosophers' different understandings of laws of nature have informed their various accounts of scientific explanation. Comparatively speaking, the unificationist model of explanation maximally fits with the regularity approach to laws of nature. (2) Scientific laws are essential in scientific explanation, so Hempel's covering law thesis is still an important dogma (doctrine) of empiricism.

5   Causal mechanism and lawful explanation

Carl Gustav Hempel proposed three models of scientific explanation: the Deductive-Nomological model (DN model), the Inductive-Statistical model (IS model), and the Deductive-Nomological model (DS model). They are also called covering law models because all scientific laws (general laws or statistical laws) are required in all three models. In Hempel's view, explanation must contain law so that law is logically prior to explanation. In other words, we can only give explanations in accordance with related law known beforehand.

With that said, Hempel's scientific explanation models have been confronted with a series of challenges. Afterward Hempel, the concept of "scientific explanation" has been further analyzed and developed by philosophers of science, among whom Wesley C. Salmon's causal/mechanical explanation model has had a profound impact. Indeed, in recent years, many philosophers of science have adopted the causal/mechanical explanation instead of the concept of law. In this chapter, we will attempt to criticize Salmon's causal/mechanical explanation model and analyze the relation between explanation, causation, and law.

Section 1   Salmon: causation and explanation

Salmon points out that Hempel's scientific explanation models falls prey to the irrelevance objection. For example, he constructs the following counter-example to the DN model:

All men taking birth control pills will not become pregnant.
John is a man taking birth control pills.

John is not pregnant.

This explanation conforms to the DN explanation model; however, we would not regard that as a good explanation because a man would never become pregnant regardless, thus rendering the information "John takes birth control pills" irrelevant and redundant.

Salmon also criticizes the requirement of high probability in Hempel's IS explanation. He points out that it is the "statistical relevance" rather than the high probability that is likely to be the key of statistical explanation, and he provides another example to demonstrate:

> Most people who get N-type psychoneurosis will recover after receiving psychiatric treatment.
> John got N-type psychoneurosis and received psychiatric treatment.
> $$\overline{} \, (r)$$
> John recovered.

According to Hempel's IS model, this explanation is a good statistical explanation if r has a high probability. However, Salmon thinks that since there is a proportion (r') of people with N-type psychoneurosis who would recover without psychiatric treatment, it is possible for John to recover automatically rather than receiving psychiatric treatment. Therefore, whether the psychiatric treatment is a relevant explanation depends on whether r is bigger than r', that is, whether the psychiatric treatment can increase the probability of recovery.

The statistical relevance Salmon raises is a kind of probability calculation, which can be defined in the following terms: under the condition A, factor C is relevant to statistic B if and only if $P(B/A\&C) \neq P(B/A)$. For example, if the psychiatric treatment changes the probability of John's healing, then it is statistically relevant. Otherwise, it is irrelevant information. Salmon further states that causality, instead of the statistical relevance, has explanatory implications.

From this, Salmon (1998) comes to five conclusions. (1) We must put the "cause, back into because." Even if some types of explanation turn out not to be causal, many explanations do appeal essentially to causes. We must build into our theory of explanation the condition that causes can explain effects but effects do not explain causes. By the same token, we must take account of temporal asymmetries; we can explain later events in terms of earlier events, but not vice versa. (2) High probability is neither necessary nor sufficient for scientific explanation. The key is whether the explanation term is statistically relevant, namely, that the explanation term would increase the probability that the term being explained would appear. (3) We can dispense – as Hempel himself did (1977) – with the doctrine of essential epistemic relativity of the IS explanation. Salmon introduces the concept of the "objective homogeneous denotational set," claiming that the statistical explanation should have objective correctness just as the DN explanation does. (4) The theory of scientific explanation should leave a place for the pragmatics of explanation. (5) We should give up searching for a universally applicable model of scientific explanation and instead observe the forms of explanation in specific subjects (pp. 302–320).

In this chapter, we will provide a critique of Salmon's account based on four aspects: (1) mark transmission; (2) singular causation; (3) causation and law; (4) ontic and epistemic conceptions.

Section 2   Mark transmission

Salmon thinks causation is a physical connection and should be understood as a process rather than an event. Using the special theory of relativity, Salmon regards the process as a line in the Minkowski coordinate system and the event as a point. The process takes more time and occupies more space than the event. For instance, the movement of a ball and the propagation of light are both physical processes.

Following Hans Reichenbach, Salmon claims that a causal process can transmit marks, but a non-causal process or a pseudo process cannot. For example, a moving truck can deliver goods from one place to another, but the shadow of the truck, which also moves with the truck, cannot transmit anything itself. The speed of causal processes must be no faster than the speed of light; otherwise, there could be a reversal of cause and effect. However, the pseudo process may exceed the speed of light. For example, if a searchlight has enough intensity of light and rotational speed, and the cloud layer is high enough, the speed of the light spot emitted by the searchlight on the layer of clouds can exceed the speed of light.

Salmon (1984) gives his own definition of mark transmission, stating:

> Let P be a process that, in the absence of interactions with other processes, would remain uniform with respect to a characteristic Q, which it would manifest consistently over an interval that includes both of the space-time points A and B ($A \neq B$). Then, a mark (consisting of a modification of Q into Q'), which has been introduced into process P by means of a single local interaction at point A, is transmitted to point B if P manifests the modification Q' at B and at all stages of the process between A and B without additional interventions.
>
> (p. 148)

In Salmon's view, there are two standards for the causal process: being able to transmit its own structure and being able to transmit the change of the structure. Concerning this issue, J. Woodward (1989) points out that the two standards are conceptually separated (pp. 375–376). For example, a photon can keep its inertia of motion, which can be regarded as a causal process according to the first standard. However, if the structure of the photon – if there is any – cannot be changed, it is not able to fulfill the second standard. Salmon (1984) also gives a definition of causal interaction:

> Let P_1 and P_2 be two processes that intersect with one another at the space-time point S, which belongs to the histories of both. Let Q be a characteristic that process P_1 would exhibit throughout an interval (which includes subintervals on both sides of S in the history of P_1) if the intersection with P_2 did not occur; let R be a characteristic that process P_2 would exhibit throughout an interval (which includes subintervals on both sides of S in the history of P_2) if the intersection with P_1 did not occur. Then, the intersection of P_1 and P_2 at S constitutes a causal interaction if: (1) P_1 exhibits the characteristic Q before S, but it exhibits a modified characteristic Q' throughout an interval immediately following S; and (2) P_2 exhibits the characteristic R before S, but it exhibits a modified characteristic R' throughout an interval immediately following S.
>
> (p. 171)

However, this definition of causal interaction must rely on the concept of marking. Consider the example of a car that is bent out of shape after hitting a tree and has cause the tree to fall down. We would usually regard this as a causal interaction. But even the shadow of the car and the shadow of the tree also meet at the spot S of time and space, followed by the shadow of the car changing in shape and the shadow of the tree moving when the tree falls, we would never think there is a causal interaction between the shadow of the car and the shadow of the tree. Conversely, however, the concept of marking, which is also causal, must also rely on the concept of causation. Hence, S. Psillos (2002) points out that since Salmon uses mark transmission as the condition distinguishing the causal process from the pseudo process, although the concept itself is causal, it still encounters the problem of having a circular definition (pp. 116–120).

Section 3   Singular causation

Salmon accepts singular causation and prefers causal realism. The two female philosophers G.E.M Anscombe and Nancy Cartwright have had much discussion on this subject, and they both think the casual verbs and causal concepts in ordinary language justify the reality of causation. For example, Anscombe writes:

> The word "cause" can be added to a language in which are already represented many causal concepts. A small selection: scrape, push, wet,

carry, eat, burn, knock over, keep off, squash, make (e.g., noises, paper boats), hurt. But if we care to imagine languages in which no special causal concepts are represented, then no description of the use of a word in such languages will be able to present it as meaning cause.

(as cited in Psillos, 2002, p. 71)

Cartwright also thinks that the relation between the substantial causal verb (e.g., *break* and *lick*) and the general concept of causation is the relation between concrete and abstract. For example, when we observe "breaking the vase," we know the causal relation directly from observation instead of through reasoning.

However, the author thinks singular causation can be identified only in the form of regularity. For example, the following sentences only have their verbs changed:

(1) John *knocks* the vase, and then it breaks to pieces.
(2) John *blows* on the vase, and then it breaks to pieces.
(3) John *looks* at the vase, and then it breaks to pieces.

We cannot find a big difference between the three verbs simply through linguistic analysis, but the cause of the vase's breaking is quite different. In sentence (1), knocking is the cause of the vase breaking; in sentence (2), however, we can hardly identify the cause because it is also possible that the vase had cracks beforehand, and the blowing plays only a part in the breaking of the vase; finally, in sentence (3), we can be sure that looking is not the cause at all, for looking is a process of light reflecting from the vase to the eyes, and this has no causal impact on the vase.

Therefore, whether or not a verb is regarded as a causal verb depends on our understanding of scientific laws. A change in scientific knowledge can even change our conception of causation. For example, in the early stages of the development of optics, some natural philosophers thought sight is the result of eyes casting light upon objects. According to that knowledge background, looking could potentially be the cause of the vase breaking, for the light may exert some force or pressure on the cracks in the vase. However, modern science tells us that sight is a process in which objects reflect light to the eyes, so looking has no causal effect on an object. Therefore, causation is not purely objective, as it also depends on humans' knowledge background.

Section 4   Causation and laws

Salmon thinks that causation is a kind of "at-at theory" – that is, causation takes effect at some place at some point in time, which need not be described by laws.

The author has criticized the concept of singular causation in an earlier section and thinks that causation can be expressed by laws, at least regularly. Here we will discuss in detail whether there are some scientific laws that cannot be expressed in the form of causation. In fact, Salmon himself hints at such a possibility, frankly acknowledging that his causal mechanism account is severely challenged by the identical particles correlation in quantum mechanics.

In their jointly written 1935 paper titled "Can Quantum-Mechanical Description of Physical Reality be Considered Complete?" A. Einstein, B. Podolsky, and N. Rosen propose the famous EPR paradox. Later, John Bell put forward Bell's theorem and designed a series of experiments that could be used to test the EPR paradox. Alain Aspect's experiments in 1982 further showed that if a pair of photons are radiated by an excited atom, the measuring result of one photon can affect the result of the other. The experimental result can be described with a law: the pair of photons or particles are in the entangled state, and their physical quantities are equal or opposite.

However, the result cannot be formulated in the form of causation. If interference on the measurement of one particle causes a change of the physical quantities of the other particle, it must be due to the former particle transmitting some mark to the other particle. The speed of mark transmission between two particles can exceed the speed of light, thus producing the action-at-a distance, which is strictly forbidden by the theory of relativity.

Therefore, the author thinks that some laws in physics cannot be expressed in the form of causation, yet causation can be expressed in the form of laws. So, the concept of scientific law should be more basic and broader than that of causation. Employing scientific law rather than causation is thus more suited to the developmental trend of modern science.

Section 5   Ontic and epistemic conceptions

Salmon (1989) summarizes three conceptions of scientific explanation, which can even be traced back to Aristotle: (1) the epistemic conception, following Hempel's approach holds that to explain something is to explicate its nomic expectability, but do further correction and modification – for example, van Fraassen's pragmatics of explanation and Philip Kitcher's unificationist model. (2) The modal conception sees explanation as showing that the explanandum must happen – for example, Mellor's modal interpretation of probability. (3) The ontic conception sees explanation as the discovery of the causation and internal mechanisms of phenomena, as well as showing its place in the broader picture of nature and its hierarchical structure – for example, A. Coffa, Salmon, and Railton (pp. 120–122).

Salmon thinks scientific explanation should be ontic, so his image of explanation is bottom up. In contrast, Kitcher suggests that scientific explanation is top down and aims to give a global understanding of superficial phenomena. Hence, he proposes the concept of the explanatory store. As he sees it, for any phenomenon K to be explained, if $E(K)$ is the argument set which best unify K, then $E(K)$ is the explanatory store of K.

Here we can take Newtonian mechanics as an example: many phenomena are in demand of explanation, such as the movements of the planets, the rhythm of tides, and the falling of objects. The three laws of motion and the law of gravity in Newtonian mechanics constitute the schematic sentences, and the mathematical method of Newtonian mechanics gives the "argument pattern." Thus, all the massive and complicated phenomena can be unified with Newtonian mechanics. In this sense, the system of Newtonian mechanics constitutes the explanatory store of the phenomena of object movement. This sheds light on Kitcher's slogan of scientific explanation as "only connect."

Salmon calls for peaceful coexistence between Kitcher's top-down approach and his bottom-up approach, claiming they express two different but compatible aspects of scientific explanation. However, the author thinks Salmon may confuse the ontology and epistemology of explanation. For example, epistemologically we can know a friend before knowing his parents, but ontologically speaking, his parents must exist before his birth. In this way, ontology and epistemology are two different aspects.

The author thinks that scientific research is a type of epistemic activity performed by human beings, and thus belongs to the category of epistemology. Ontologically, without the existence of humans, cats still eat the mice, and tides rise and fall. However, without human beings, cats cannot explain to mice that the appearance of cats causes the disappearance of mice, and the tides do not themselves recognize that the gravitational pull of the Moon can explain their rise and fall. Therefore, scientific explanation is a kind of epistemic activity of human beings and belongs to the category of epistemology.

Salmon himself stresses that the theory of scientific explanation should provide a place for the pragmatics of explanation, so it is safe to assume he would also agree that scientific explanation belongs to epistemic activities, since pragmatics only belong to epistemic activities, and there is no such thing as pragmatics in the purely natural world. If, in this way, we admit that scientific explanation is a kind of epistemic activity of human beings, then it should be Kitcher's top-down approach, rather than Salmon's bottom-up approach.

Section 6   Preliminary exploration of relations among law, causation, and explanation

Law, causation, and explanation are all central but controversial concepts in the general philosophy of science. A short essay is certainly not able to address all of the relations between the three concepts. This chapter, by criticizing Salmon's causal mechanism account, simply tries to argue that law, as opposed to causation, is essential for scientific explanation.

The author still insists on Hempel's covering law thesis and thinks that the law should be the most basic concept in scientific explanation, with even causation being understood in the form of scientific laws. In comparison, Kitcher's top-down approach is fully compatible with the MRL view of scientific laws and thus may generate the least debate. The MRL view, which was proposed by Mill, Ramsey, and Lewis, is the view that scientific knowledge provides us with the best deductive system to understand the world. The axioms and theorems of the system constitute scientific laws, while other universally true statements are accidental generalizations. Scientific knowledge, as a deductive system, provides a global understanding of various phenomena in the world. Moreover, the MRL view of scientific laws and Kitcher's unificationist model of scientific explanation are perfectly connected.

The author tends to think that ontological events and processes, including causal events and causal processes, come first, and it is from these that we induce the scientific laws and form our scientific knowledge system. However, epistemologically, only with the background of scientific knowledge can we identify some regularities as genuine laws and then use those laws to explain and identify the causal link between events. Thus, this chapter proposes that the ontological order should be: events (or processes), regularities (including genuine laws), and the scientific knowledge system. In contrast, the epistemological order should be: the scientific knowledge system, scientific laws, and explanation and causation.

If demonstrated with graphs, the ontological structure is represented by the left picture, with the direction being bottom up; the epistemological structure is seen in the picture on the right, with the direction being top down.

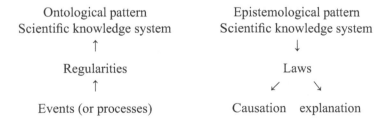

Ontological pattern   Epistemological pattern
Scientific knowledge system   Scientific knowledge system
↑   ↓

Regularities   Laws
↑   ↙   ↘

Events (or processes)   Causation   explanation

6 Is there such a thing as a *ceteris paribus* law?

Section 1 The rise of *ceteris paribus*

The author is a defender of the notion that laws of nature play an essential role in explanation and causation. In this chapter, I will discuss a recently active topic in the philosophy of science: the so-called *ceteris paribus* law. The use of *ceteris paribus* (hereafter CP) clauses in philosophy and sciences has a long history. According to Persky (1990), the idea of CP in economics can be traced back to William Petty's *Treatise of Taxes and Contributions* (1662), and Alfred Marshall used the terminology for the first time in his *Principles of Economics* (1890) (Earman, Roberts, and Smith, 2002, p. 1).

Most generalizations of economics usually hold true only in "the absence of disturbing causes" or with "other things being equal." Thus, John Cairnes, in his *Some Leading Principles of Political Economy Newly Expounded* (1874), gives a classic example of a CP law: "The rate of wage, other things being equal, varies inversely with the supply of labour." As seen in this example, the CP clause is usually read as "other things being equal," "there are no interferences," "in the absence of disturbing factors," or the like. With that said, proponents of the CP law do not think it possible to make the CP clause explicit, because there could be an infinite number of conditions. The CP law has also been phrased as "If CP, then all *F*s are *G*s."[1]

In recent years, the topic of CP laws has become an important issue in the philosophy of science, particularly in the discussion of the status of special sciences such as biology, geology, economics, and psychology, which are different from fundamental sciences. According to the standard account, science studies the laws of nature that should be strict, that is, universal and exceptionless. However, most generalizations of the special sciences are not so. This being the case, how can we still justify their scientific status?

Consider a generalization in economics: "When the demand for a product increases, while supply remains constant, the price of the product will increase" only holds true if other things are equal. Many interferences, such as irrational behavior and ignorance on the part of customers, could bring about counteracting results. Likewise, in biology, Mendel's law of segregation would be invalid if there had been certain changes in the initial conditions of the Earth or in the evolutionary chain. So, then, how can we still insist on the scientific legitimacy of the special sciences?

In philosophy of science, there are mainly three views regarding CP generalizations. The first view maintains that the laws of the special sciences are CP laws but that CP laws are just as scientifically legitimate as the strict laws of fundamental sciences. Thus, the special sciences are scientific even without strict laws.

The second view goes further. Some philosophers suggest CP all the way down, and they propose that the majority of the laws of physics are CP laws; in this way, for these philosophers, there is no fundamental difference between fundamental and special sciences. For example, Peter Lipton (1999) states, "Most laws are *ceteris paribus* (CP) laws" (p. 155). Similarly, as Michael Morreau (1999) argues, "hedged laws are the only ones we can hope to find. Laws are commonly supposed to be truths, but interesting generalizations, without some modifier such as '*ceteris paribus*,' are by and large false" (p. 163).

The third view is more traditional and is supported by John Earman, John Roberts, and Sheldon Smith (hereafter ERS) in a paper titled "*Ceteris Paribus* Lost?" They believe only fundamental laws in physics should be considered laws; in contrast, for them, CP generalizations are not laws at all. For these scholars, there is no law in the special sciences, although the special sciences are still scientifically legitimate.

At this point, it is worth noting that the current edition of the *Stanford Encyclopedia of Philosophy* includes an entry on *ceteris paribus* laws written by Reutlinger, Schurz, and Hüttemann (2011). This fact alone indicates a growing interest in the philosophical discussion of CP laws and their critical roles in contemporary sciences. For this reason, assessing the various views on CP laws is a justified and timely task.

The author generally agrees with the second view of CP laws (see above), but, instead of expounding a defense of this view, this chapter will focus on criticizing the approach of the third view as represented by ERS. The goal is to validate CP laws as scientific laws insofar as they apply to special sciences such as biology, economics, and sociology.

Section 2   *Ceteris paribus* lost?

John Earman[2] is a distinguished University Professor Emeritus in the Department of History and Philosophy of Science at the University of Pittsburgh, specializing in the history and philosophy of physics, as well as the general philosophy of science. He received a PhD degree from Princeton University in 1968 and then taught at UCLA, Rockefeller University, and the University of Minnesota. He is a fellow of the American Academy of Arts and Sciences and served as the President of the Philosophy of Science Association (2000–2001). In his paper titled *"Ceteris Paribus* Lost?" Earman – along with co-authors Roberts and Smith – has launched perhaps the fiercest attacks against CP laws. With the contention that there should be no room for CP laws in physics, these arguments can be characterized as follows:

(i)   *Appealing to examples taken from physics.* It seems to ERS that in order for a "real" CP law to be interesting, its CP clause must be ineliminable, and the range of this clause cannot be made explicit. As they state: "Otherwise, the CP clause is merely a function of laziness: Though we *could* eliminate the CP clause in favor of a precise, known conditional, we choose not to do so" (Earman, Roberts, and Smith, 2002, pp. 283–284). They argue that CP clauses can easily be eliminated by known conditions if we use scientific language properly.

(ii) *Confusing Hempel's provisos with ceteris paribus clauses.* ERS think that Hempel's central concern "is not the alleged need to save law statements from falsity by hedging them with CP clauses, but rather the problem of applying to a concrete physical system. . . . [T]he conditions of the provisos are conditions for *the validity of the application*, not conditions for *the truth of the law statements of the theory*" (Earman, Roberts, and Smith, 2002, p. 285). Thus, they would accept Hempel's provisos but reject CP clauses.

(iii) *Confusing laws with differential equations of evolution type.*[3] ERS propose a distinction between a theory consisting of a set of non-hedged laws and the application of a theory that might be hedged (in an easily stateable way). They argue that those examples provided by CP law proponents are just differential equations of evolution type: "But differential equations of evolution type are not laws; rather, they represent Hempel's applications of a theory to a specific case. They are derived using (unhedged) laws along with non-nomic modeling assumptions that fit (often only approximately) the specific case one is modelling. Because they depend on such non-nomic assumptions, they are not laws" (Earman, Roberts, and Smith, 2002, p. 286).

(iv) *Opposing Cartwright's view of component forces.* ERS raise two objections to Nancy Cartwright's local antirealism about component forces on two counts: (a) in many cases the component forces are measurable; (b) it is logically unclear to conclude that something is not occurrent just because it is not measurable (Earman, Roberts, and Smith, 2002, p. 287). Thus, ERS appear to propose a local realism about component forces to support their (local) realism about the (component) laws.

(v)   *Opposing Cartwright's argument from Aristotelian natures and the modern experimental method.* ERS argue that Cartwright's primary goal is to negate a "Humean" view that confines the ontology of science to the behaviors of physical systems and regularities thereof, and favors a broader ontology that includes natures and capacities. In order to emphasize that laws can be linked to (supervene on) behaviors even without the concept of capacity, ERS reiterate their supervenience thesis in this part: "One can grant that there is a lot more to being a law of nature than just being a true behavioral regularity, and even grant that what laws state is helpfully understood in terms of capacities, while maintaining that laws (and capacities) must supervene on the behaviors of physical systems" (Earman, Roberts, and Smith, 2002, p. 288).

(vi) *The world being a messy place.* The proponents of CP laws would argue: "The world is an extremely complicated place. Therefore, we have no good reason to believe that there are any non-trivial contingent regularities that are strictly true throughout space and time." ERS do not accept this as a valid inference. They acknowledge the premise of this argument as "undeniably true," but they deny the conclusion.

In addition, ERS raise two objections to the CP laws of sciences other than physics: (1) that there seems to be no acceptable account of their semantics; and (2) that there seems to be no acceptable account how they can be tested. ERS go on to say that the first objection is "not fatal to CP laws," while the latter objection of untestability deals a decisive blow to CP laws. They argue:

> [E]ither this [CP] auxiliary can be stated in a form that allows us to check whether it is true, or it can't. If it can, then the original CP law can be turned into a strict law by substituting the testable auxiliary for the CP clause. If it can't, then the prediction relies on an auxiliary hypothesis that cannot be tested itself. But it is generally, and rightly, presumed that auxiliary hypotheses must be testable in principle if they are to be used in an honest test. Hence, we can't reply on a putative CP law to make any predictions about what will be observed, or about the probability that something will be observed. If we can't do that, then it seems that we can't subject the putative CP law to any kind of empirical test.
>
> (Earman, Roberts, and Smith, 2002, p. 293)

In summary, ERS have given three main theses in their argument against CP laws:

(1) CP clauses can be easily eliminated if we use scientific language properly – that is, ERS's (i).

(2) There are no ways to test CP laws if we cannot substitute testable auxiliaries for the CP clauses.

(3) The so-called CP laws are just differential equations of evolution type (which are hedged on non-nomic assumptions); thus, they are by no means laws – that is, ERS's (ii), (iii), (iv), and perhaps (vi).

I will argue against the first and second theses in Sections 3 and 4, respectively. The third thesis is more convoluted, as it involves a deep understanding of the laws of nature; therefore, I will discuss it in detail in Section 5.

Section 3   Eliminability

ERS insist on the eliminability of CP conditions. In other words, they claim that if we use scientific language properly, CP clauses can be easily eliminated by known conditions. If, in this way, CP conditions could be eliminated by hedged conditions in scientific language, then the so-called CP laws would precisely confine the scope of application for those laws of nature. Thus, to defend the CP laws, we must argue for the ineliminability of CP conditions.

Marc Lange, who is a supporter of CP laws, raises an example to this point: to state the law of thermal expansion (the change in length of an expanding metal bar is directly proportional to the change in temperature),

> one would need to specify not only that no one is hammering the bar on one end, but also that the bar is not encased on four of its six sides in a rigid material that will not yield as the bar is heated, and so on.
>
> (Lange, 1993, p. 234)

To counter, ERS argue this example is expressed in a language that "purposely avoids terminology from physics." If we use technical terms from physics, the conditions can be easily stated as, "The 'law' of thermal expansion is rigorously true if there are no external boundary stresses on the bar throughout the process" (Earman, Roberts, and Smith, 2002, p. 284). Nevertheless, how can we be sure any forces on the metal bar, such as gravity of the Earth and electric force from electric charges nearby, would not be stresses that can influence the expansion of the metal bar? Even if we agree with ERS's strict terminology, suppose the temperature of the bar is raised higher than its melting point; in this case, would the length of the metal bar still be proportional to its temperature? In fact, upon further analysis, we find that ERS do not mention the melting temperature at all in their rigorous reconstruction of the thermal expansion law.

ERS give another example to bolster their claims about eliminability: "Kepler's 'law' that planets travel in ellipses is only rigorously true if there is no force on the orbiting body other than the force of gravity from the dominant body and vice versa" (Earman, Roberts, and Smith, 2002, p. 284). But we might ask, is "other than" standard terminology in physics?[4] Even if it is, would the ellipse law still hold true if, for example, the mass of the Sun increased or decreased dramatically because of a certain nuclear reaction – which is not a "force" at all – inside the Sun? If we consider all such interferences, ERS's rigorous reformulation may finally have to expand infinitely.[5]

Section 4   Untestability

Are CP laws testable? ERS quote two common views regarding the testability of CP laws: (1) we can confirm the putative law that "CP, all *F*s are *G*s" by finding evidence that in a large and interesting population, *F* and *G* have a highly positive correlation; (2) we can confirm the hypothesis that "CP, all *F*s are *G*s" if we find an independent, non-ad-hoc way to explain away every apparent counter-instance, that is, every F that is not a *G*.

ERS think the former view lends confirmation to the stronger claim that in some broader classes of populations, *F* and *G* are positively correlated, which would not be a CP law. The latter view, in their estimation, is not sufficient. Here is their counter-example: "CP, white substances (or compounds containing hydrogen) are safe for human consumption." As they see it, although we may be able to explain away why any particular white substance is not in fact safe for human consumption, whether with explanations from modern biology or medicine, the CP statement itself is not a law at all.

ERS's two objections are excellent, especially the first. However, it may be possible to find another kind of testability for CP laws: namely, that their contrapositions are testable.[6] If we write a CP law "If CP, then L" as "CP $\to$ L", it is logically equivalent to "$\neg$L $\to$ $\neg$CP." Framed this way, we can logically ascertain its contraposition: "If not all *F*s are *G*s, then not CP."

With this in mind, we might here consider two interpretations of the CP condition, "there are no interferences" and "other things being equal." The former has the logical form $\neg(I_1 \lor I_2 \lor I_3 \ldots)$, with *Ii* referring to different Interferences, which could be an infinite set. The latter can be written as $(E_1 \land E_2 \land E_3 \ldots)$, with *Ei* here meaning various equal conditions, which can also be infinite. Therefore, whenever *F* is not *G*, there must be at least some interference or an unequal condition. Thus, experimenters try to find the interference or unequal condition. If they finally find one or the other, it becomes clear that there is either a disturbing factor or something unequal.[7] That should be a confirmation of the CP law!

As for ERS's "white substances" counter-example, it involves the distinction of genuine laws and accidental generalizations, which continues to be an actively discussed topic in the philosophy of science. For example, both "All gold spheres are less than 1 mile in diameter" and "All uranium spheres are less than 1 mile in diameter" are true universal propositions, but the former is merely an accidental generalization, while the latter is a law of nature due to uranium's critical mass. Though the very nature of laws of nature is philosophically controversial (for details see Section 5), it is important to point out that we can still distinguish laws of nature from accidental generalizations.

Similarly, even if the generalization "CP, white substances (or compounds containing hydrogen) are safe for human consumption" is true, we regard it as an accidental generalization rather than a genuine law because this CP generalization appears to have no connection with our best knowledge from modern science. However, Mendel's principle of segregation is a CP law because it has been an important part of modern genetics. Thus, we see that ERS's counter-example can be solved by looking to the distinction between accidental generalization and genuine law.[8]

Section 5   Differential equations and supervenience

ERS argue that those so-called CP laws, such as thermal expansion law and Kepler's law, are merely differential equations of evolution type. They are not laws at all. However, such a claim may be inconsistent with Earman's own so-called system approach (Carroll, 2004, p. 2) to the understanding of laws of nature, so this inconsistency deserves further discussion.

In the philosophy of science, there are mainly two camps in the debate about laws of nature. Scholars like David Armstrong, Michael Tooley, and Fred Dretske insist on a necessitarian view, arguing that a kind of physical or nomic necessity distinguishes genuine laws from accidental generalizations. With that said, it is quite difficult for them to work out an explicit definition of necessity. Therefore, nowadays the opposing views of scholars such as J. S. Mill, Frank Ramsey, and David Lewis (i.e., MRL) are more popular in philosophical circles. John Earman belongs to the MRL camp.

MRL regard laws as "consequences of those propositions which we should take as axioms if we knew everything and organized it as simply as possible in a deductive system" (Ramsey, 1978, p. 38). Thus, "a contingent generalization is a law of nature if and only if it appears as a theorem (or axiom) in each of the true deductive systems that achieves a best combination of simplicity and strength" (Lewis, 1973, p. 73). Accordingly, for MRL and Earman, "All uranium spheres are less than 1 mile in diameter" is a law of nature, while "All gold spheres are less than 1 mile in diameter" is only an accidental generalization. This is because the former belongs to the knowledge system of quantum physics, whereas the latter does not.

The system approach (of MRL and Earman) acknowledges not only axioms, but also theorems of our deductive system in the modern sciences as laws of nature. For instance, we might think of such cases as the thermal expansion law or Kepler's law, which is the consequence of modern physics – more precisely, solid mechanics – or we might also consider the law of universal gravitation. If Earman maintains these are only differential equations of evolution type, his view of CP laws may be inconsistent with ERS's claim regarding laws of nature.

ERS insist on the distinction between conditions for the truth of a law (CP) and conditions for the validity of its application (provisos). They accept Hempel's notion of provisos, but reject the CP clause. Conceptually, conditions for the truth of a statement may not be equivalent to the conditions for the validity of its application. For example, the statement "Independence Day in the USA is July 4th" is definitely true, but it is not so applicable when pursuing girls. Granted, the situation in fundamental physics is a bit different.

The fundamental laws of physics are always abstract, and on account of this, there is no direct way for us to justify their truth. That certainly does not mean that those fundamental laws are untestable. Indeed, we can derive some testable effect, usually with the help of bridge principles (or correspondence sentences), from the abstract laws combined with non-nomic assumptions or initial conditions. That provides us with a kind of application of abstract laws to real situations, and since their derivations are testable, we can confirm, or disconfirm, the truth of the fundamental laws.

Thus, the conditions for the truth of a law are closely related, though not logically equivalent, to their applications. Consider ERS's supervenience thesis in which they maintain that laws of nature must supervene on the behaviors of physical systems. According to the *Oxford English Dictionary*, *supervene* means "to come on or occur as something additional or extraneous; to come directly or shortly after something else, either as a consequence of it or in contrast with it; to follow closely upon some other occurrence or condition." Therefore, the supervenience thesis reminds us that laws always "come after" real behaviors.

Consider a nice analogy raised by Cartwright. She regards the relationship between laws of nature and real situations as a kind of abstract–concrete relationship, somewhat like morals and fables. To support this, she quotes G. E. Leesing's claim, "The general exists only in the particular and can only become graphic (anschauend) in the particular." To demonstrate what she means by morals, we might say that the weaker are always prey to the stronger; then, to put flesh on this abstract notion, we can consider a real and concrete situation, such this one described in Lessing's fable: "A marten eats the grouse. A fox throttles the marten; the tooth of the wolf, the fox" (Cartwright, 1999, pp. 37–43). Cartwright's analogy is a wonderful example of supervenience: the moral supervenes on the fable.[9]

In what sense, then, is the moral true? The moral provides a nice idealization of the real situation, such as the relationship between martens, grouses, foxes, and wolves, yet it can also be applied to the relationships between other animals, and perhaps even to relations among humans or nations. However, if we consider a possible world in which there are no animals, humans, or nations, the moral need would no longer be true, for there is nothing upon which it can supervene. Now consider another possible world in which the electric charge of all things is removed – perhaps because of a certain evolution of the universe – while other things (i.e., laws) remain equal. Will the law of electromagnetic force, one of the four fundamental forces in modern physics, still hold true in such a possible world? According to his own distinction between the conditions for truth and for application, Earman can argue the law of electromagnetic force still holds even though it no longer applies to anything. However, according to his supervenience thesis, the law cannot supervene on any physical behavior since there is no electric charge. Again, it is quite doubtful that Earman's view of CP laws is inconsistent with his claim regarding laws of nature.

A possible reason why ERS stick to the strictness of laws is the mathematical tradition itself, which continues to manifest a kind of residual Platonic idealism within the modern sciences. According to Thomas Kuhn (1977), there have been two traditions – mathematical versus experimental – in the development of physical science. Because in ancient Greece the classical sciences, astronomy, harmonics, mathematics, (geometrical) optics, and statics were all practiced by a single group and situated in a shared mathematical tradition, Kuhn also deems the mathematical tradition a "classical mode." In contrast, the experimental tradition, on account of resulting mainly from the work of Francis Bacon, is a "Baconian mode." As Kuhn sees it, "Into the nineteenth century the two clusters, classical and Baconian, remained distinct" (Kuhn, 1977, p. 48).[10] Although both traditions can be found in Europe, the center for Baconian work was Britain, and for the mathematical, the Continent, especially France. While, since the advent of the 20th century, the long-standing division between mathematical and experimental traditions has been more and more obscured, even seemingly disappearing in some respects, Kuhn maintains that "it continues to provide a source of both individual and professional tensions" (Kuhn, 1977, p. 64).

Interestingly, Cartwright also mentions the distinction between the English and French traditions. She quotes Pierre Duhem's two types of thinking: the deep but narrow-mindedness of the French, and the broad but shallow-mindedness of the English. She quite humorously states her personal sympathy for the "English" type of thinking: "The realist thinks that the creator of the universe worked like a French mathematician. But I think that God has the untidy mind of the English" (Cartwright, 1983, p. 19).

At the same time, Alexander Koyre (1978) proposes that Galileo's scientific revolution is a revival of Platonism, although Galileo himself preferred to be regarded as "New Archimedes." Regardless, while it may be crude to name the mathematic tradition as a Platonic mode, the modern sciences do appear to unconsciously assume a kind of Platonic view. Plato's central doctrines of idealism are well known: eternal, changeless, and paradigmatic ideas are more real and perfect, but the world that appears to our senses is defective and wretched (Kraut, 2013). Similarly, many scientists and philosophers maintain that laws of nature are true in an ideal realm. On this view, the laws of nature, at least after discovery and confirmation, should be true forever, no matter what actually happens in the world. If a real observation does not conform to the laws of nature or their derivation, it is simply because the real situation is not as ideal as required. Those scientists and philosophers regard reality as a copy – usually imperfect – of the ideality, and such a view appears to be a remnant of Platonic idealism.

The author does not intend to argue against Platonic idealism in this book, but it is important to note that there is some tension between the Platonic view and the Humean view (especially the supervenience thesis) of laws of nature. If Earman wants to be a coherent Humean, why not reverse the order, instead considering the laws of nature as the idealization – so often imperfect – of our real world?

Section 6 Summary

Earman and his colleagues have raised several objections to the *ceteris paribus* laws. Contra Earman, in this chapter, the author argues that CP clauses can be ineliminable even if we use scientific terminology properly and that it is also possible to test the contraposition of a CP law – and, thus, the law itself. This chapter also points out that Earman's account of differential equations may be inconsistent with his MRL view of laws of nature. Additionally, the author suggests there may be a kind of residual Platonic idealism in the modern sciences, probably because of the mathematical tradition, and the Platonic view of laws of nature may be inconsistent with Earman's supervenience thesis, or perhaps with his Humean empiricism. However, if we give up the remnant of Platonic idealism in science and philosophy, the concept of CP laws will not be as difficult to accept.

Notes

1 The author is not satisfied with this formulation. Dr. Liying Zhang, at the Central University of Finance and Economics in Beijing, China, is currently working on the logic of CP statements using generic logic, which shall be a promising approach.

2 The author learned a great deal from Professor John Earman during time spent as a visiting fellow in the Center for Philosophy of Science at the University of Pittsburgh in the 2005–2006 academic year. As perhaps the greatest sign of

respect one can afford another scholar is to study his or her subject of interest and to try one's best to raise some serious criticism, the author wishes to express the utmost gratitude and respect for Prof. Earman with this chapter.

3 For example, the acceleration of free-falling objects equals g, which can serve as an example of a differential equation because it can be deduced from Universal Gravitation Law and some non-nomic assumptions, such as the diameter and mass of the Earth, and the assumption that there is no air friction.

4 If there is no clear-cut distinction between the language of physics and ordinary language, perhaps because they always overlap, it seems unnecessary for ERS to persist in stressing "terminology from physics."

5 Here, ERS provide two reasons why there is no CP law: (a) the CP clause is easily eliminable by a known condition, and (b) they are not laws anyway (but differential equations) (Earman, Roberts, and Smith, 2002, p. 284). The author will argue against the second claim in Section 5.

6 The author learned this idea from Mehmet Elgin and Elliot Sober (2002). They raise a contraposition argument against Nancy Cartwright's claim that fundamental laws do not apply in the real world.

7 If they do not find either, that shall serve as a kind of practical falsifiability of a CP law, though falsifiability-in-principle of a law will always meet the problem posed by the Duhem-Quine Thesis.

8 It seems to be possible to work out the distinction between a genuine CP law and an accidental CP generalization by MRL's view, which will be discussed in Section 5. Here is a simplified example: Consider a CP law, "CP $\rightarrow L_1$", which is an axiom of our best deductive system. From the axiom, with other initial conditions O, we can logically get CP $\rightarrow (L_1 \wedge O)$. If the consequent implies another lawlike sentence L_2, then "CP1 $\rightarrow L_2$" can be a theorem from the system, rendering another CP law.

9 Here we need not persist in antirealism about the fundamental laws, for the moral "The weaker are always prey to the stronger" can still be true in the idealization of the real situation. Cartwright (1999, p. 23) writes, "Nowadays I think that I was deluded about the enemy: it is not *realism* but *fundamentalism* that we need to combat."

10 Galileo and Newton are apparent exceptions, but Kuhn acknowledges only Newton as a real exception. While Galileo's dominant attitude toward that aspect of science remained within the classical mode, Newton did participate unequivocally in both traditions. For instance, his *Principia* lies squarely within the tradition of the classical science, whereas *Opticks* is in the Baconian (Kuhn, 1977, pp. 49–50).

7 Explanation and reduction

Explanation and *reduction* are two closely related concepts. For instance, Ernest Nagel has used Hempel's scientific explanation model to represent his own theory of reduction. With that said, *reductionism* is a widely criticized term in contemporary philosophical circles. However, since the concept of reduction has various interpretations, reductionism could also be understood in various ways. Because of this, many of the controversies regarding reductionism and anti-reductionism are due to different understandings of the concept of reduction. This chapter first attempts to analyze and clarify the varying conceptions of reduction, and then provides some brief remarks on different versions of corresponding reductionism. The goal is that, after this discussion, both reductionist and antireductionist scholars will be able to express more clearly their notions of reduction and views about reductionism.

Section 1   Language reduction

In science and philosophy, reductionism has a long history. However, systematic analyses of reductionism were not attempted until the rise of the philosophy of science. In 1929, Hans Hahn, Rudolf Carnap, and Otto Neurath, with the assistance of F. Waismann and H. Feigl et al., published the famous programmatic manifesto "The Scientific Conception of the World: The Vienna Circle." This manifesto not only marks the founding of logical positivism as represented by the Vienna Circle, but it is also widely regarded as the beginning of "the Unity of Science Movement" and physicalism. "The unity of science" treats all sciences – for example, biology, psychology, social science – as unified. The Vienna Circle believes that these disciplines can ultimately be reduced to physics in a general sense, a view that is, consequently, also named physicalism. Therefore, many view this manifesto as a program of reductionism.

In this manifesto, the concept of *reduce* (including various expressions such as *reduction, reduced, reducibility, reducible, reductive*, etc.) occurs many times. According to the present author's own calculation, the total occurrences amount to 11 times and can be divided into four usages: (1) to be reduced to language entities (e.g., statement or concept), which occurs 4 times; (2) to be reduced to experiences (or the given, intuition), which occurs 4 times; (3) to be reduced to a state of affairs or relation, which occurs twice; (4) to be reduced to laws, which occurs only once (Hahn et al., 1929, pp. 321–340).

Hence, the author tends to think "the unity of science" places a strong emphasis on the linguistic unity of all disciplines; physicalism also focuses on the reducibility of all discipline languages to the physical language. Of course, because the manifesto was written in tandem by numerous philosophers, there might be some inconsistencies between those scholars on these points. Nevertheless, if we look at the academic leader of the Vienna Circle Rudolf Carnap's individual works from a later period, the notion of language reduction will become clearer.

Carnap holds a discreet attitude toward the unity of laws, writing:

> Thus there is at present no unity of laws. The construction of one homogeneous system of laws for the whole of science is an aim for the future development of science. This aim cannot be shown to be unattainable. But we do not, of course, know whether it will ever be reached.
>
> (Carnap, 1938, p. 403)

However, he firmly argues for the unity of language. He states:

> On the other hand, there is a unity of language in science, viz., a common reduction basis for the terms of all branches of science, this basis consisting of a very narrow and homogeneous class of terms of the physical thing-language. This unity of terms is indeed less far-reaching and effective than the unity of laws would be, but it is a necessary preliminary condition for the unity of laws. We can endeavor to develop science more and more in the direction of a unified system of laws only because we have already at present a unified language.
>
> (Carnap 1938, p. 404)

While arguing for the unity of language, Carnap distinguishes physical language from thing-language. The physical language contains, in addition to logico-mathematical terms, all and only physical terms. The thing-language, on the other hand, contains: observable thing-predicates, such as *red, blue, cold, hot* (but not *temperature*), *light*, and *heavy* (but not *mass*); disposition-predicates (reducible to observable thing-predicates), such as *transparent, fragile,* and *soluble*; substances, like *stone, water,* and *sugar*; and processes, like *rain, fire,* etc. Carnap thinks that all discipline languages can be reduced to physical language, and then physical language to thing-language, which itself can ultimately be reduced to observable thing-predicates. Hence, Carnap's notion of the unity of science is a kind of radical empiricism.

Declaring the significance of the unity of science, Carnap writes:

> And, in addition, the fact that we have this unity of language is of the greatest practical importance. The practical use of laws consists in making predictions with their help. The important fact is that very often a prediction cannot be based on our knowledge of only one branch of science. . . . For very many decisions, both in individual and in social life, we need such a prediction based upon a combined knowledge of concrete facts and general laws belonging to different branches of science. If now the terms of different branches had no logical connection between one another, such as is supplied by the homogeneous reduction basis, but were of fundamentally different character, as some philosophers believe, then it would not be possible to connect singular statements and laws of different fields in such a way as to derive predictions from them. Therefore, the unity of the language of science is the basis for the practical application of theoretical knowledge.
>
> (Carnap, 1938, p. 404)

Section 2   Micro-reduction

In 1958, P. Oppenheim and H. Putnam distinguished three types of concepts of reduction in their collaboratively written article "Unity of Science as a Working Hypothesis." In this article, they described the first kind of reduction as "theory reduction," which can be framed as follows: given two theories T_1 and T_2, T_2 is said to be reduced to T_1 if and only if (1) the vocabulary of T_2 contains some terms not in the vocabulary of T_1; (2) any observational data explainable by T_2 are explainable by T_1; (3) T_1 is at least as systematic as T_2. (Normally T_1 is more complicated than T_2.)

The second kind of reduction defined by Oppenheim and Putnam is "branch reduction" or "discipline reduction," namely the reduction of a branch of science B_2 by another branch B_1 – for example, the reduction of chemistry to physics. The necessary and sufficient conditions for branch reduction are as follows: considering the accepted theories of B_2 at a given time t as T_2, then B_2 is reduced to B_1 at time t if and only if theory T_2 of B_2 can be reduced to T_1 of B_1. Besides this, there can be partial reduction between two branches – that is, some part of theories in B_2 can be reduced to theories in B_1.

The final kind is *micro-reduction*. With this type, the conditions for the reduction of B_2 to B_1 are as follows: (1) B_2 is reduced to B_1; (2) the objects in the universe of discourse of B_2 are wholes that decompose into proper parts, all of which belong to the universe of discourse of B_1. Beyond this, micro-reductions are characterized by three properties: (1) they are transitive; (2) they are irreflexive; (3) they are asymmetric. From the viewpoint of Oppenheim and Putnam, being transitive is the most crucial of these properties for the unity of science because it implies that micro-reduction has the characteristic of accumulation.

Hence, these three concepts of reduction build upon one another, for only under the full completion of theory reduction can we obtain branch reduction, and only by adding other conditions will branch reduction be considered micro-reduction.

Both Oppenheim and Putnam (1958) advocate micro-reduction, and they propose six conditions for it:

> (1) There must be several levels; (2) The number of levels must be finite; (3) There must be a unique lowest level (i.e., a unique "beginner" under the relation 'potential micro-reducer'); This means that success at transforming all the potential micro-reductions connecting these branches into actual micro-reductions must, ipso facto, mean reduction to a single branch; (4) Anything of any level except the lowest must possess a decomposition into things belonging to the next lower level. In this sense each level, will be as it were a "common denominator" for the level immediately above it; (5) Nothing on any level should have a part on any higher level; (6) The levels must be selected in a way which is "natural" and justifiable from the standpoint of present-day empirical science. In particular, the step from any one of our reductive levels to the next lower level must correspond to what is, scientifically speaking, a crucial step in the trend toward over-all physicalistic reduction.
>
> (p. 409)

They also list six levels of micro-reduction (Oppenheim and Putnam, 1958, p. 409):

6   Social groups
5   (Multi-cellular) living things
4   Cells
3   Molecules
2   Atoms
1   Elementary particles

As Oppenheim and Putnam (1958) see it, micro-reduction is a working hypothesis. Arguing as to why this is the case, they write:

(1) It is of practical value, because it provides a good synopsis of scientific activity and of the relations among the several scientific disciplines; (2) It is, as has often been remarked, fruitful in the sense of stimulating many different kinds of scientific research. By way of contrast, belief in the irreducibility of various phenomena has yet to yield a single accepted scientific theory; (3) It corresponds methodologically to what might be called the "Democritean tendency" in science; that is, the pervasive methodological tendency to try, insofar as is possible, to explain apparently dissimilar phenomena in terms of qualitatively identical parts and their spatiotemporal relations.

(p. 413)

Micro-reduction is a strong version of reductionism, yet our current understanding of reductionism is mostly due to this concept. Still, as mentioned above, micro-reduction relies on the other two types of reduction, for if the primary reduction level – that is, theory reduction – is problematic, then it will be hard to achieve a higher level of reduction. Because of this limitation, we will focus on "theory reduction" in the next section.

Section 3   Theory reduction

It was Ernest Nagel who first gave a sophisticated explication of the concept of theory reduction. Nagel was born in Slovakia and immigrated to the United States with his family at the age of 10. He received his PhD in 1931 from Columbia University and afterwards spent his career teaching there. Because of his great contributions to the philosophy of science, he was elected as a fellow of both the National Academy of Sciences and the American Academy of Arts and Sciences. Nagel's philosophical position is close to Logical Positivism, and his monographs include *The Meaning of Reduction in Natural Sciences* (1949), *Sovereign Reason* (1954), *The Structure of Science* (1961), *Teleology Revisited and Other Essays in the Philosophy and History of Science* (1979).

By the 1960s, Hempel's scientific explanation models were already well established, and Nagel, borrowing the logical structure of scientific explanation, proposed the structure of theory reduction. For instance, we can explain Galileo's law of a freely falling body with Newton's law of universal gravitation according to Hempel's DN model:

Newton's law of universal gravitation: $F = GMm/r^2$
Initial conditions: the mass of the Earth is M,
   its radius is r, etc.   explanans

Galileo's law for a free-falling body: $S = 1/2gt^2$ explanandum

Nagel thinks such an explanation in fact constitutes a theory reduction because Galileo's law can be logically deduced from Newton's gravitational law when initial conditions are also accounted for, so Galileo's law can thus be reduced to Newton's law.

Nagel carefully distinguishes between two kinds of reduction, with one being homogeneous reduction, or the notion that a reduced law either can be deduced from explanatory premises or is the approximation of a law deduced from explanatory premises. This is pertinent to scientific practice because simplification and approximation are quite common in theory reduction. For instance, if we consider the example of Galileo's law being reduced to the law of gravity, we usually presuppose the Earth is a perfect sphere, that the falling body is evenly forced, and that said body's change in height from the Earth can be ignored when comparing it with the radius of the Earth. In addition, only by ignoring air resistance can we finally derive Galileo's law approximately from the law of gravity.

In a homogeneous reduction, the reducing law and the reduced law use the same vocabulary (such as mass, length, time, force, etc.). If the reducing law contains some vocabulary not found in the reduced law, however, then it might be an inhomogeneous reduction, which is the second kind of reduction according to Nagel. For instance, thermodynamics can be reduced to statistical mechanics, but statistical mechanics lacks some thermodynamic concepts, such as temperature, pressure, etc. Likewise, some of chemical theory can also be roughly reduced to quantum theory in physical chemistry; nevertheless, quantum theory lacks some chemical concepts, such as valence. This being the case, how then should we deal with this type of reduction?

To provide an answer, Nagel provides two conditions of theory reduction: (1) the condition of derivability – that is, that T_2 can be reduced to T_1 means all the statements of T_2 are derivable from that of T_1; (2) the condition of connectibility – that is, if T_1 does not have contain any non-logical expressions of T_2, then the vocabulary of T_1 can be connected with the various words of T_2 simply by appending certain additional premises (Nagel, 1974, pp. 907–915).

Hence, through the bridges law or correspondence rules, the vocabulary of the reduced theory can be connected with that of the reducing theory. For example, the concept of temperature in thermodynamics is connected with the average kinetic energy of molecules in statistical mechanics; similarly, pressure is connected with the average value of molecules exerting a force on the wall of a vessel. As for chemical valence, it can be connected with the outer electron distribution of elements in quantum theory.

As for Nagel's impact, his analysis of theory reduction has come to form a central part of the classical literature of logical positivism, and it is for this reason that Paul Feyerabend chooses it as a target in his criticism of the concept of reduction in empiricism. Feyerabend (1963) thinks empiricists' theory of reduction demands two conditions: (1) the consistency condition, for only such theories are admissible in a given domain that either contain the theories already used in this domain, or that are at least consistent with them inside the domain; and (2) the condition of meaning invariance, for meanings should be invariant with respect to scientific progress – that is, all future theories should be phrased in such a manner that their use in explanations does not affect what is said by theories or what factual reports are to be explained.

Following Kuhn's theory of incommensurability between paradigms, however, Feyerabend refutes both conditions. First, the consistency condition is extremely intolerable, because it excludes a theory on the grounds of it being contradictory with other theories rather than facts. Second, if we attempt to explain scientific theories in the way accepted by scientific circles, then most correspondence rules are either wrong or meaningless. This is because if they affirm the existence of an entity that is negated by theory, they will be wrong, but if they assume its existence, they will be meaningless (Feyerabend, 1963, pp. 926–932).

Hence, Feyerabend thinks an old theory is replaced by a new theory, rather than being reduced to a new theory. For instance, Newtonian mechanics cannot be reduced to the theory of relativity; instead, the theory of relativity replaced Newtonian mechanics. Indeed, if we admit the thesis of incommensurability between paradigms, Feyerabend's criticism about theory reduction seems to be quite persuasive. However, in scientific practice, most scientists are reluctant to accept such a strong version of incommensurability. Thus, we would do better to develop arguments, whether for or against theory reduction, within science itself.

Section 4　Counter-examples in biology

Traditionally, philosophy of science has regarded physics as the paradigmatic discipline among the various sciences. In recent years, with the development of biology, however, philosophy of science has increasingly turned to biology for inspiration. For example, in the *Stanford Encyclopedia of Philosophy*, there are three entries about reductionism: "scientific reductionism" in the general philosophy of science, "Reductionism in Biology" in the philosophy of biology, and "Intertheory Relations in Physics" in the philosophy of physics (table of contents).

In the entry "Reductionism in Biology," Ingo Brigandt and Alan Love define three types of reductionism: (1) ontological reduction, or the notion that each particular biological system (e.g., an organism) is constituted by nothing but molecules and their interactions – also called "compositional materialism"; (2) methodological reduction, or the notion that biological systems are most fruitfully investigated at the lowest possible level and that experimental studies should be aimed at uncovering molecular and biochemical causes – also called "decomposition strategy"; and (3) epistemic reduction, or the notion that knowledge about one scientific domain (typically about higher level processes) can be reduced to another body of scientific knowledge (typically concerning a lower and more fundamental level), for example, theory reduction and explanatory reduction (Brigandt and Love, 2008, first section).

These days, biological circles have reached a consensus to accept ontological reduction. In the 1930s, there were still quarrels about vitalism, but now biologists widely believe that life is ultimately constituted of nothing but molecules and their interactions and that there is nothing mysterious in the living world.

Methodological reduction has also increasingly become a dominant and guiding concept within biological research. Since the publication of Erwin Schrödinger's famous book *What Is Life*, life scientists have used more and more physical methods to study biology. Moreover, molecular biology even occupies both the frontier and the mainstream of current biological research. Hence, in biology, most disputes about reductionism are mainly with respect to theory reduction and explanatory reduction.

In biology, the dispute about theory reduction focuses on whether or not traditional biology can be reduced to molecular biology. On this point, Philip Kitcher (1984) takes genetics as his example to argue against theory reduction. Kitcher thinks that in order for classical genetics to be reduced to molecular biology, three conditions must be satisfied: (R1) classical genetics contains general laws about the transmission of genes, which can serve as the conclusions of reductive derivations; (R2) the distinctive vocabulary of classical genetics can be linked to the vocabulary of molecular biology by bridge principles; and (R3) a derivation of general principles regarding the transmission of genes from principles of molecular biology would explain why the laws of gene transmission hold. Kitcher then argues against all three of these theses.

First, taking Mendel's Law as example, Kitcher points out that it cannot be represented as a first-order logical law in scientific practice, as it is not universally true; thus, we should instead treat it as an inference skill or an explanatory pattern. Moreover, Nagel's condition of derivability might be invalid as well, since Kitcher suspects that Mendelian genetics can be derived from molecular biology.

Second, according to the condition of connectibility, if there are differences between the empirical (non-logical) vocabulary of two theories, bridge laws are required to link them together. Hence, we need a definition like the following logical expressions (in which M is a molecular gene):

$\forall x(X$ is a classical gene $\leftrightarrow Mx)$

However, the so-called molecular gene is only a physical structure, such as the length of DNA, the number of nucleotide pairs, a codon, etc. Therefore, it is doubtful whether the entity in pure physical language is identical to Mendel's gene.

On this point, Kitcher proposes an argument appealing to the multi-realization of function-structure because, in biology, the correspondence between function and structure is often in the form of many-to-many. In other words, at the molecular level, there is no single structure corresponding to a particular kind of function. For example, both insects and birds have the function of a wing, but their physical structures are completely different. The same molecular gene in a different context may have a variant genetic expression. In this way, we observe a many-to-many correspondence between physical structures and functional genes, and this reality fails to meet the requirement of the condition of connectibility.

Finally, Kitcher thinks molecular biology provides no explanation of Mendelian genetics, for, as he puts it, we do not need "gory details" to achieve an explanation. Take, for example, why the distribution of gene pairs in non-homologous chromosomes is mutually independent, a fact for which traditional cytology provides us with the following answer: in the stage of meiotic division, chromosomes are aligned in pairs according to homology. Homologous chromosomes can exchange genetic materials and produce chromosome recombinants. In meiotic division, each recombinant distributes its members to a respective gamete; hence, the distribution of the members in different recombinants might be mutually independent. This is already a satisfying answer, so we need not know of what the chromosome consists–that is, the gory details at the molecular level (Kitcher, 1984, pp. 971–1003).

With the rise of the concept of emergence in recent years, anti-reductionism has gradually prevailed in the philosophy of biology. For example, Sandra Mitchell explicitly asserts that "emergence" in biology makes epistemic reduction untenable, pointing out three characteristics of emergence to support this assertion: (1) novelty, or the notion that the whole often possesses some properties that are absent in its parts – that is, "the whole is greater than the sum of its parts"; (2) unpredictability, which entails that it is hard to accurately predict the system's behaviors because the system is recursive and non-linear, with the individual interactions within said system forming intricate relationships; (3) downward causation, which is the idea that the whole can influence its parts through positive or negative feedback mechanisms. These three characteristics of emergence, as Mitchell (2008) argues, prevent systems biology from being reduced to molecular biology.

Section 5   Explanatory reduction

Now that we have discussed theory reduction in biology, we can continue to analyze explanatory reduction, which concerns whether or not macro-explanation can be reduced to micro-explanation. For example, if we want to explain the deaths of rabbits in a particular area (mainly hunted by foxes), can we simply use the micro-event "rabbit r is eaten by fox f at time t" to explain the macro-event? To give a macro-explanation here is to explain the deaths of the rabbits, whereas micro-explanation explains that rabbit r is eaten by fox f in place p at time t. With this difference in mind, can macro-explanation be reduced to micro-explanation?

Alan Garfinkel (1981) thinks micro-explanation is different from macro-explanation. In his view, these two types of explanation actually answer different questions; thus, the latter cannot be reduced to the former, as demonstrated below:

The object of the macro-explanation: why the rabbit was $\left\{ \begin{array}{l} \text{eaten} \\ \text{not eaten} \end{array} \right\}$

The object of the micro-explanation: why the rabbit was eaten $\left\{ \begin{array}{l} \text{by fox at time } t \dots \\ \text{by some other fox} \dots \end{array} \right\}$

Framed this way, we can consider another example. If we want to provide a macro-explanation for the deaths of rabbits, we may say an excessive number of foxes is restricting the population of the rabbits. In contrast, a micro-explanation may say rabbit r comes out at time t and happens to meet fox f, thus being eaten by the fox. If the rabbit had not come out at time t, according to micro-explanation, then it would not have been eaten by fox f. However, these micro-explanations cannot provide us with the genuine macro-explanation: even if rabbit r is not eaten by fox f at time t, so long as the number of foxes is large, the rabbit will still have a high probability of being eaten by another fox.

Subsequently, Garfinkel thinks macro-explanation is superior to micro-explanation on two counts: (1) in terms of practical application, if we would like to eradicate or prevent something, micro-explanation is sometimes useless for rendering a solution. (2) Explanation should tell us what could have been otherwise, but micro-explanation is too sensitive to conditions, which may cause it to meet the problem of redundant causality (Garfinkel, 1981, pp. 445–448). He maintains,

> The idea is that no matter what the substratum turns out to be, we can proceed independently to construct upper-level explanations. So far, this is a fairly modest claim. . . . I want to make a stronger claim: that in many cases, the micro-level is inadequate, and we therefore must construct upper-level explanations.
>
> (p. 451)

In conclusion, macro-explanation cannot necessarily be reduced to micro-explanation, meaning explanatory reduction is also problematic.

Section 6   Summary

In this chapter, we have discussed the relation between explanation and reduction and, in doing so, have explored a series of concepts such as language reduction, theory reduction, discipline reduction, micro-reduction, ontological reduction, methodological reduction, and explanatory reduction. With this explication complete, we can now summarize and remark upon these types of reduction.

Language reduction attempts to unify all discipline languages by physical language – and, ultimately, by thing-language. The author thinks unifying scientific languages is of practical significance, but if we confine language reduction to Carnap's "reduction sentence," then, according to the development of logical empiricists' criterion of cognitive significance, language reductionism will face the problem of holism. Our scientific language is constructed as a partially interpreted system, which can be used to explain experience, but we need not reduce the scientific language to experience sentence by sentence. Still, if we understand language reductionism as a kind of radical empiricism, which assumes science contains no mysterious content beyond experience, then it can still have positive values.

Theory reduction means the reduced law can either be deduced from explanatory premises or be the approximation of the explanandum. Discipline reduction means all the laws in a particular discipline can be reduced to the laws in another discipline. Micro-reduction requires not only discipline reduction, but also that the objects in the domain of the discipline, as a whole, can be decomposed into proper parts all of which belong to the domain of another discipline. Theory reductionism, however, is challenged by the problem of incommensurability between paradigms, as well as a whole host of difficulties and counter-examples from biology, so it is also problematic. Consequently, discipline reductionism and micro-reductionism, which are both based on theory reductionism in one way or another, are both seen to be improbable.

In biology, ontological reduction usually means each particular biological system (such as an organism) is assumed to be constituted of nothing but molecules and their interactions. This view is also called "compositional materialism." Currently, ontological reductionism is widely accepted in both scientific and philosophical circles, and it is also a weaker version of reductionism. This is because, at least up to this point, it is difficult to imagine – much less argue with certainty – that biological and mental phenomena are composed of anything other than physical materials.

Methodological reduction means that biological systems are most fruitfully investigated at the lowest possible level and that experimental studies should be aimed at uncovering molecular and biochemical causes. This is also called "decomposition strategy." At its current stage, methodological reductionism is replete with practical significance. However, we cannot and should not ignore mutual learning between disciplines. For example, biology should certainly learn from the mathematical and experimental methods of physics, while, at the same time, the tradition of natural history in biology might merit physicists' attention as well.

Epistemic reduction means the knowledge about one scientific domain (typically at a higher level) can be reduced to another body of scientific knowledge (typically concerning a lower and more fundamental level), and it includes theory reduction and explanatory reduction. As we have revealed, theory reductionism can be seen as infeasible, especially in biology. Explanatory reduction suggests that macro-explanation be deduced from micro-explanation, and according to Garfinkel's analyses, it is also problematic. Since these two kinds of epistemic reduction – theory reduction and explanatory reduction – are both fraught with many difficulties, epistemic reductionism should also be viewed as infeasible as a result.

8 Scientific explanation and historical interpretation

Section 1 The application of scientific explanation models in historical studies

Many people believe that Hempel's scientific explanation models can only be applied to the natural sciences – that is, they would claim that scientific explanation is necessary only in a narrow sense; however, in the humanities and social sciences, only interpretation can be used, which means scientific explanation is no longer applicable. This chapter will mainly deal with this question, and after doing so, will then analyze the similarities and differences between scientific explanation and historical interpretation. Finally, it will explore the methodological unification of the natural sciences and the humanities.[1]

After exploring his scientific explanation models and their modified forms, Hempel thinks that explanations in history also fit his models because historical explanations require general laws as well. With that said, these laws and initial conditions are not very accurate, as they are sometimes either so ambiguous or so trivial that they only appear in the form of a partial explanation or an explanation sketch. He particularly raises two examples, "genetic explanation" and "rational explanation," which have been common in historical explanations, illustrating that both of them also fit his scientific explanation models (Hempel, 1968, pp. 68–79).

A genetic explanation provides an explanation for one certain event by using a kind of narrative form, which describes the entire developmental process of the event. For instance, the occurrence of event D can be traced back to event A initially, moving from there to B, then to event C, and finally to event D, and this sequence constitutes the entire developmental procedure. That is, an account of the process from event A to event D provides a narrative explanation of the final historical event D.

Hempel claims this kind of genetic explanation fits his scientific explanation models because a genetic explanation begins with the first stage of the narrative history, which then continues to the second stage. The occurrence of the second stage, which bears a nomological relation with the first stage, can therefore be explained by the characteristics of the first stage. Similarly, the characteristics of the second stage can be used to explain the third stage. That means a genetic explanation, in fact, is a series of explanations of separated stages: event A explains event B, event B explains event C, and finally event C explains event D. The whole process is subject to Hempel's explanation models.

A rational explanation is usually used to explain somebody's motivated action in history, and it is framed by philosopher of history W. Dray as follows:

A was in situation C;
It is appropriate to do X in situation C;
Thus, A did X.

Hempel, however, argues that's rational explanation only explains that A should do X, but it cannot explain the fact that A actually did X. In light of this, Hempel revises Dray's rational explanation model to formulate the following version:

A was in situation C;
A wants to act rationally;
Any rational people in situation C will, or are highly probable to, do X;
Thus, A did X.[2]

The revised rational explanation also fits with Hempel's scientific explanation, leading him to think that his explanation models not only can be applied within the natural sciences, but also can be extended to human sciences. He proudly claims: "In so doing, our schemata exhibit, I think, one important aspect of the methodological unity of all empirical science" (Hempel, 1968, p. 79).

Section 2 Deductive thesis and causal explanation?

Many philosophers criticize the deductive thesis of Hempel's scientific explanation models, which is usually related to the notion of causal explanation. Alan Donagan tries to show that the deductive thesis is necessary in scientific explanation models and then argues causal explanation cannot be applied to historical studies (Donagan, 1959, p. 430). Yuankang Shih, an Emeritus Professor at the Chinese University of Hong Kong, explicates this argument even further.[3] As he thinks, if we want to explain the occurrence of event E, we must remove the possibility that event E may not happen at all. For example, to explain "A building in Hong Kong is on fire at some time," we obviously cannot use "An ant dies in Australia at some time" to provide an appropriate explanation. The reason for this is that the statement "An ant dies in Australia at some time" cannot remove the possibility that "A building in Hong Kong is on fire at some time" would not happen. However, the deductive thesis of scientific explanation reveals that, according to deductive logic, the explanandum can be logically deduced from the explanans. This ensures that the expanandum is necessarily true when the explanans is true, thus removing the possibility that the explanandum would not happen.

Subsequently, Yuankang Shih (1983) believes that to explain the cause of an event is to uncover the sufficient conditions for that event; furthermore, he combines explanation together with cause, writing: "one way to make an explanation is to find out the causes of the event" (p. 23). However, in the IS model, the inference from the explanans to the explandum is actually inductive, rather than a logical deduction. Thus, Yuankang Shih argues that the deductive thesis and the IS model are in conflict with each other, and he points out: "with the acceptance of the Inductive-Statistical Model, we cannot insist this thesis any more. Meanwhile, due to the Inductive-Statistic Model, positivists have to modify their understandings of explanation drastically" (Shih, 1983, p. 102).

Donagan and Shih both identify scientific explanation with causal explanation, and then they argue that historical events cannot be explained with concepts such as cause or causal law, but instead with concepts like reason or intention. In fact, this kind of criticism of Hempel's explanation models is quite common in the philosophy of history; nevertheless, the author disagrees with their opinion, as such a criticism seems to show a misunderstanding of the concept of explanation in science. It is true that causal explanation is a kind of scientific explanation, but not vice versa. Causal explanation is one form of scientific explanation, but it cannot cover all kinds of scientific explanation.

First of all, causal explanation can only reveal the connection between different events, while scientific explanation not only can reveal the connection between different events, but also the connection between different laws. Because of this, Hempel (1965) claims:

> [C]ausal explanation in its various degrees of explicitness and precision is not, however, the only mode of explanation on which the D-N model has a bearing. For example, the explanation of a general law by deductive subsumption under theoretical principles is clearly not an explanation by causes.
>
> (p. 352)

For example, the law of gravity can explain the law of free-falling bodies, but the explanation is not a causal explanation.[4]

Another point that Hempel does not discuss in detail but that is generally accepted by most scientists is that causal laws used in causal explanations are deterministic laws, whereas the laws used in scientific explanations can be either deterministic laws or statistical laws.

With the development of quantum mechanics in the 20th century, scientists discovered that micro-particles can only be described in the form of a probability wave, which follows statistical laws rather than deterministic laws. As a result, Carnap (1995) declares that "the determinism in the 19th century has been abandoned by modern physics" (p. 288). Accordingly, he suggests redefining causal laws with the use of scientific laws – that is, any causality in the world can be expressed with scientific laws. If we want to study causality, we must do research on these laws, which includes showing how they are expressed, confirmed, or falsified by experiments (Carnap, 1995, p. 227).

We can even conclude that deterministic law, after all, is simply a special case of statistical law because, according to the discovery of quantum mechanics, macro-phenomena following deterministic laws are fundamentally made up of micro-phenomena following statistical laws. For example, Ayer (1956) thinks that a causal law can be regarded as a statistical law whose probability is 100% and is, thus, the "limiting case" of statistical laws (p. 816).

While it did make some sense to define scientific explanation as causal explanation in the 19th century, with the development of contemporary physics in the 20th century, such understanding is no longer correct. Many historians and philosophers insist that causal explanation cannot be applied in historical studies, yet when they criticize Hempel, they certainly misunderstand the notion of scientific explanation in the philosophy of science.

With the similarities and differences between scientific explanation and causal explanation thus classified, we may now assert that Hempel's scientific explanation models may inherit the mathematical tradition in the natural sciences. In other words, the inference from explanans to explanandum is a kind of mathematical calculation, which is not always a logical deduction. In the DN model, the inevitable validity of the explanandum can be calculated from the explanans, and the same is true with the DS model; in the IS model, however, the explanandum can also be calculated from the explanans with a high probability, so the explanandum is approximately true too.

Thus, in scientific explanation models, the covering law thesis is the most important, while the deductive thesis is not essential. Therefore, it is in conflict with the IS model. Hempel (1965) calls his DN, IS, and DS models the covering law models, because all that the models must cover contains scientific laws (p. 412).

Section 3 Covering law thesis and history?

Following the discussion in Section 2, the central problem is no longer whether causal explanations can be applied in historical studies or whether the deductive thesis is in conflict with the IS model; instead, the main concern is whether or not there must be laws in historical explanations.

German philosopher Wilhelm Windelband once suggested that history is an "idiographic science," while science is a "nomothetic science" (Windelband, 1980, pp. 165–168). But does idiographic science need general laws? Both Donagan and Dray say no, and Shih (1983) presents their criticism of Hempel very carefully in his article "Positivism and Historical Explanations" (pp. 104–108).

Donagan (1959) argues that humans have free will, so there cannot be any universal law in human events. For instance, to answer the question of why government organizations continuously expand, Hempel uses three general laws:

(1) People who have jobs are never willing to lose their jobs.
(2) When people have adjusted to some skills, they are not willing to make any changes.
(3) People will never want to lose the power they have, while they will still want greater power.

Evidently, the above three laws all can be met with counter-examples: somebody may prefer to retire earlier and give up his or her job; somebody would like to change his or her skillset; though many people love power, there are also hermits who dislike power. Historical events involve too many human factors, so Donagan believes there are no universal laws in historical events, and, as a result, scientific explanation cannot be applied in historical studies.

Nonetheless, Donagan's position that "there is no law at all in human events" seems to be too absolute, for if we can find even one general law in human events, we can falsify his opinion. Although people have free will, which leads to their free choice, there can be regularities resulting from collective actions, which may have nothing to do with individuals. For example, if the Marxist theory that "Productive forces determine production relations" is true, it is a social law at the macro level. Although individuals have their own choice, they cannot affect the validity of the macro law.

In addition, we have classified the distinction between scientific explanation and causal explanation, so if there are statistical laws in the historical field, those historical explanations can satisfy the covering law thesis. For example, although Donagan worries about the counter-examples to the three economic laws, they can be transformed into three statistical laws that are highly probable and, therefore, can explain the expansion of government. The three new laws are as follows:

(1) People who have good jobs are usually unwilling to lose their jobs.
(2) When people have adjusted to some skills, they are quite probably unwilling to make any changes.
(3) People who will never want to lose the power they have had and only want greater power account for a high proportion of all people.

Dray does not deny there can be laws in history, yet he thinks that historical events are all so special that scientific explanation cannot be applied to them. For example, explaining why Louis XIV was unpopular in later years, many historians may mention a series of wrongheaded policies he put into effect, such as exhausting all the resources to develop military power, the persecution of pagans, corruption of the royal court, etc.; however, if we, like a logician, summarize the situation as a proposition that "If any governor in the same situation that Louis XIV once experienced carries out all of Louis XIV's policies $P_1, P_2, P_3. \ldots$, then he or she will be unpopular," would this be a scientific explanation of history? We can frame the explanation this way:

If any governor in the situation that Louis XIV once experienced carries out all of Louis XIV's policies $P_1, P_2, P_3. \ldots$, then he or she will be unpopular.   (General law)

Louis XIV was in that situation.   (Initial condition)

Louis XIV was unpopular in later years.   (Explanandum)

Dray claims that this kind of law is too specific, for it is impossible to find another Louis XIV in the world, meaning there would be no other instance that could satisfy the law. After all, how can a general law only be applied in one special case? In this way, he denies the role of law in historical explanation.

As Hempel distinguishes initial conditions from scientific laws in his explanation models, we can try to defend Hempel from this aspect. Dray's specific law can be reformulated as "Any governor may increase the probability of his unpopularity if he or she carries out some policies," especially when these are a collection of bad policies.[5] To answer the question "Why was Louis XIV unpopular?" we can take the policies carried out by Louis XIV, such as P_1, P_2, P_3, as initial conditions, and then combine them with the above reformulated law to explain Louis XIV's unpopularity. Certainly, we cannot precisely calculate the probability of how those policies contributed Louis XIV's unpopularity, but we can frame the issue as follows:

Any governor may increase the probability of his unpopularity if he or she carries out some policies.   (Statistic law)

Louis XIV carried out policies like P_1, P_2, P_3. . .   (Initial condition)

Louis XIV was unpopular in later years.   (Explanandum)

The modified argument can be regarded as an "explanation sketch," providing directions and a scheme for the explanation of "Louis XIV was unpopular in later years." In the explanation sketch, although the initial conditions about Louis XIV are unique, it does not mean the related historical laws are also unique. Because of this, we can successfully avoid Dray's criticisms.

In summary, although Donagan and Dray both question the covering law thesis of scientific explanation models is valid in historical studies, they cannot deny that there may be general laws, at least statistic laws, in historical studies. So Hempel's scientific explanation models can be valid in humanities and social sciences too.

Section 4   Meaningful behaviors and objectivity?

British philosopher Peter Winch thinks it is impossible to apply methods of the natural sciences to the humanities due to the objectivity problem of the social sciences. Winch is influenced deeply by the late Wittgenstein, so he uses Wittgenstein's arguments on language and entity, as well as his concepts such as "language game," to deny the validity of the methods of the natural sciences being applied in the social sciences. One of the reasons he gives is that human action is rule governed and can be characterized as "meaningful behavior" (Winch, 1990).

There is no doubt that human action is influenced by cultural rules and norms, which makes many explanations in the social sciences rely on these rules and norms. However, is a rule explanation the same as a scientific explanation using scientific laws (including deterministic laws and statistical laws)?

In ancient Greece, scholars made distinctions between "truth by nature" and "truth by convention," with the former being universally valid and the latter varying by culture. Physics is a typical kind of truth by nature because it is regarded as inherent in nature, and every culture must follow physical laws. Language, on the contrary, represents a kind of truth by convention. For instance, Chinese and English have their own conventions, so Chinese grammar and spelling rules need not be applied in English, and vice versa.

Human action is governed by cultural rules and norms, so the explanations of those actions require normative truths. For example, automobiles move on the right side in mainland China in accordance with Chinese traffic regulations. This, however, is merely a social norm, not a general law, because in Hong Kong and Britain, automobiles move on the left side. This example of rule explanation can be formulated as follows:

Automobiles move on the left side in Hong Kong in accordance with its
    traffic regulations.   (Rule)

Someone drives in Hong Kong following the
    traffic regulations.   (Initial condition)

Thus, he or she drives on the left side.   (Explanandum)

In this explanation, we use only rules but no scientific law. So, it seems that Hempel's covering law is useless to explain human actions, which implies that Hempel's scientific explanation models are not applicable within the human sciences.

However, we can try to defend Hempel's covering law thesis once again, this time by setting the cultural rules and norms as initial conditions and then adding some scientific laws. Hence, the above rule explanation can be transformed into the following form:

It is the safest to drive on divided roads according to proscribed
    directions.   (Scientific law)

According to the safety principle, Hong Kong has established
    the traffic regulation that automobiles should move on the
    left side.   (Initial condition)

Someone drives in Hong Kong following the
    traffic regulations.   (Initial condition)

Thus, he or she drives on the left side.   (Explanandum)

This kind of explanation conforms to Hempel's scientific explanation models, so Winch and other scholars' assertion that "human actions are governed by rules" may choose another target: the objectivity problem of the social sciences (Shih, 1982, pp. 1–5). Natural sciences are usually considered to be objective, while the social sciences are not considered objective at all, so how can methods of the natural sciences be applied in the social sciences?

The objectivity problem of the social sciences and the humanities is a huge topic, so I can only discuss it here very briefly. In historical studies, Leopold von Ranke advocates that history should "reconstruct the past based on facts," namely, scientific history is the objective description of the past. This kind of account is called "objectivism." In contrast, many scholars emphasize that historical studies are selective and evaluative, conditions both of which depend on value. "Objectivity" must be value free, so historical studies cannot be objective. Their account is called "relativism," the main arguments of which include two aspects: first, historical topics are value-charged; second, historians are value-guided when they construct historical topics. The two aspects above constitute "the fountainhead of relativism" (Dray, 1964, pp. 23–24).

Relativists' doubts regarding the objectivity of humanities and social sciences are quite reasonable, but are natural sciences really as objective as they are claimed to be? Many philosophers and historians think that the objectivity of natural sciences is manifested by the real descriptions of natural phenomena. For example, R. G. Collingwood (1948), a famous British philosopher and historian, writes:

> To the scientist, nature is always and merely a "phenomenon", not in the sense of being defective in reality, but in the sense of being a spectacle presented to his intelligent observation; whereas the events of history are never mere phenomena, never mere spectacles for contemplation, but things which the historian looks, not at, but through, to discern the thought within them.
>
> (p. 214)

However, this viewpoint is challenged seriously by Kuhn's concept of paradigm and incommensurability. In his famous book *The Structure of Scientific Revolutions*, Kuhn (1970) divides the development of science into two parts: normal science and scientific revolution. The so-called normal science is made up of scientists' "puzzle-solving" activities under the paradigm of the scientific community. Different paradigms mean different worldviews, and the scientists in different paradigms seem to live in different worlds.

Thus, natural sciences' descriptions of the world are not inevitable and exclusive because they are relative to paradigms. Natural scientists do not "see" the world; instead they "see as." Using the Duck–Rabbit Illusion as an example, some people "see as" a duck, while others "see as" a rabbit;

similarly, scientists operating within different paradigms may also "see as" differently. For example, classical mechanics "sees" space and time "as" stationary and absolute; but the theory of relativity "sees as" the parameters of object movement and relative to its reference system. Therefore, in this sense, natural scientists are not "looking at" the nature but instead "look through" it (Kuhn, 1970).

Moreover, many philosophers think the natural sciences are value free, so they are objective. However, many other philosophers of science also question this opinion.

For instance, in Kuhn's article "Objectivity, Value Judgment, and Theory Choice," he suggests that theory choice in sciences does not rely on objectivity but instead depends on the decisions made by scientists based on their values, which include accuracy, consistency, scope, simplicity, and fruitfulness. Theory choice in the sciences is not objective, though it is neither individual nor subjective; it is "the collective judgment of scientists" who are trained in the scientific community. Furthermore, Kuhn (1977) thinks "objectivity" should be analyzed by value criteria such as accuracy, consistency, etc., which do "not imply the limitation of objectivity, but the meaning of objectivity" (pp. 320–339).

Hempel proposes a similar approach in his article "Science and Human Value." He thinks science cannot provide validity proof of "categorical value judgments," but scientific knowledge presupposes values (Hempel, 1998, pp. 110–127).

Winch points out that in the social sciences, human actions are governed by cultural rules and norms, and on account of this he questions the objectivity problem of the social sciences. However, if the natural sciences face the same problem and rely on value judgment of the scientific community too, we can suspect that the objectivity problem cannot justify the methodological separation between the natural sciences and the social sciences.[6]

The research objects of the natural sciences and the social sciences are different. The natural sciences study natural phenomena, while the humanities and social sciences study "meaningful behaviors." Even so, the object difference can only reveal that there may be some ontological differences between natural sciences and social sciences. Nonetheless, any argument that such would lead to methodological disunity is in demand of further justification. For example, biology usually studies life phenomena, while physics studies nonliving phenomena, but we would not claim there is a methodological disunity between the two disciplines.

Section 5   Explanation and interpretation?

Comparatively, Charles Taylor, a famous Canadian political philosopher, raises an even more sophisticated and serious challenge. He makes a distinction between interpretation in the humanities and explanation in the natural sciences, and he tries to separate the "science of interpretation" as a "science of man" from the natural sciences.

The philosophical meaning of the concept of interpretation can be traced back to the hermeneutics tradition founded by German philosopher Friedrich Schleiermacher. Since then, many continental philosophers including Wilhelm Dilthey, Martin Heidegger, Hans-George Gadamer, and Paul Ricoeur have discussed this topic in detail. According to R. E. Palmer (1969), hermeneutics tries to go beyond the "subject–object schema" of modern sciences, criticizing the "scientific objectivity" of the natural sciences (pp. 223–241).

Taylor thinks that the target of interpretation is always text or something like it, which can be of meaning, and it must always satisfy three conditions: first, it must have sense; second, people can separate its sense and its presentation; finally, its sense is specific to one certain subject. The intention of interpretation is to reveal the coherence of and the sense behind the target.

Taylor thinks the human being is a "self-interpreting animal" and that human actions are meaningful, which means they can only be studied with the method of hermeneutics. Furthermore, sense is relative to subject, can be separated from its presentation, and is meaningful only in some field and relative to the senses of other things (Taylor, 1998, pp. 110–127). This is apparently different from the seeking of universal law in the natural sciences.

Taylor's distinction between interpretation and explanation is very insightful. If the distinction is valid, it seems that explanation in the natural sciences and interpretation in the human sciences do not follow the same methodological pattern. However, does this imply methodological separation between the natural sciences and the social sciences?

The author thinks Kuhn's response to Taylor answers this question convincingly (Kuhn, 1998, pp. 128–134). Taylor contends that the natural sciences are universally true – or in the hermeneutic term, being of scientific objectivity – so the knowledge of astronomic phenomena is universal, which means human beings have a "heaven for all." Kuhn refutes Taylor's claim and points out that the heavenly bodies of the ancient Greeks are very different from that of modern people, as they use different classifications. The ancient Greeks classify heavenly bodies into three categories: star, planet, and meteor. Unlike modern people, they put the Sun and the Moon in the category of planet because they think the Sun and the Moon are more similar to planets such as Mars, Mercury, and Venus. Today, in contrast, we regard the Sun as a star and the Moon as a satellite. Thus, using Kuhn's terminology, knowledge within the natural sciences is relative to paradigms and is not universally valid in every culture.

In addition to paradigms, Kuhn raises the concept of the scientific community in his book *The Structure of Scientific Revolutions*. In his opinion, all concepts – no matter in the natural sciences or the humanities – are owned by communities, and there can be conceptual differences between communities because of their differences in culture or language. Thus, the concept vocabulary of the natural sciences, as well as the humanities and social sciences, may have taken on a different meaning for a different community due to the paradigm difference.

Kuhn argues there are both explanations and interpretations in the natural sciences, as well as in the humanities and social sciences. For instance, there are many interpretations in the natural sciences. For instance, the freshman students who have just joined the scientific community want to know senior scientists' interpretations of scientific symbols and the usages of scientific instruments. Especially during periods of scientific revolution, many new concepts or instruments demand new interpretations, such as a consideration of what is the meaning of "wave-particle duality" in a new paradigm or how the Nuclear Magnetic Resonance spectrometer should be used. These cannot be explained by general laws because people do not understand the symbols in those laws yet. In other words, symbols, meanings, and applications of general laws cannot be explained by the general laws themselves, but demand interpretation.

Consider these two questions from the natural sciences, which seem to have the same form:

(1) Why does the mass of a moving object, with high velocity close to the speed of light, increase?
(2) Why is time relative to the reference system?

In fact, the first question is about scientific explanation, and its answer can be derived from the special theory of relativity, whereas the second one involves our understanding of the concept of time, which cannot be explained by the theory of relativity itself but, rather, by interpretation of what is time within the paradigm of the theory of relativity.

In addition to interpretation, is there explanation within the humanities? According to Kuhn, contemporary humanities and social sciences are still in the stage of "pre-paradigm," so many schools of thought are flourishing and contending with one another. Nevertheless, if human sciences can establish their own paradigms like natural sciences, then human scientists will do the same puzzle-solving research as natural scientists. When they use laws of human sciences to elaborate human actions, those laws constitute explanations in the humanities and social sciences.

For example, the following two questions in historical studies can be classified into the categories of interpretation and of explanation:

(3) Why is the French Revolution an important event in modern history?
(4) Why did the French Revolution happen?

The third question involves how we understand the meaning of the French Revolution, which is in demand of historians' interpretations according to their value criteria. The fourth question needs historians' detailed descriptions of the initial conditions contributing to the occurrence of the French Revolution. If these historians share the same paradigm and belief the same historical laws, their explanations should be the same. However, in reality, historical studies are still in the period of "pre-paradigm," so historians from different schools will provide different explanations.

Thus, Kuhn agrees with the distinction between explanation and interpretation, but he disagrees that this difference constitutes a distinction between the natural sciences and the human sciences, too. In the author's opinion, if natural sciences have many interpretations as do the human sciences, and if the human sciences – after setting up their own paradigm – can provide explanations as do the natural sciences, there can still be methodological unity between the natural sciences and the human sciences.

Section 6   Summary

Today, there are mainly two attitudes towards methodological unity between the natural sciences and the human sciences. One is scientism, which insists that the methodology of the natural sciences can be completely extended to the human sciences. Logical positivism is the representative of scientism. In the same camp, Carnap and O. Neurath's Unity of Science movement or physicalism seeks to unify all the empirical sciences and finally reduce them to physics. However, the other attitude, in opposition to scientism, is called "idealism" by Dray (1957, p. 8). Idealists tend to believe there is an essential distinction between the human sciences and the natural sciences, so both sides should stay in their own domain and never invade that of the other.

This chapter reviews Hempel's scientific explanation models and tackles the possibility of their application in the human sciences. According to Hempel, all empirical sciences, including natural sciences and human sciences, are methodologically unified, despite their different contents. His scientific explanation models not only can be applied in the natural sciences, but also can be extended to historical studies. To demonstrate this, Hempel proposes a series of notions such as elliptic explanation, partial explanation, and the explanation sketch. In addition, he analyzes the concept of genetic explanations and rational explanations, which are both quite common in historical studies, and argues that they also fit his scientific explanation models.

Many historians and philosophers present their own disagreements, which can be summarized into four questions or concerns: the deductive thesis and causal explanation, the covering law thesis and history, meaningful behaviors and objectivity, and explanation and interpretation. We can now respond to each of these respectively. The author believes, first, that the deductive thesis is not fundamental to scientific explanation, while the covering law thesis is essential. Equating scientific explanation with causal explanation is inaccurate. Second, there are laws in historical studies, and the covering law thesis is necessary for historical explanations. Third, natural sciences and human sciences both require value judgments of academic communities and both do not "see" the world, but rather "see as." Finally, all empirical sciences involve both interpretation and explanation; hence, the distinction between interpretation and explanation cannot justify the assertion of methodological disunity between the natural sciences and the human sciences.

Of course, the discussion so far is not sufficient to prove the methodological unity of natural science and humanity, but the author agrees with Kant that pursuing the "unity of system" is one of the human ideals. Although "ideals" do not have the same "objective validity" as categories, it is a driving force to guide human actions. Just like human beings' ideal to continuously pursue the "groundwork of/for the metaphysics of morals," we will continuously pursue the unity of the natural sciences and the human sciences. Surely, the final unity may not be achieved by the natural sciences unifying the human sciences, as early logic positivists suggested. However, with the realization of more human factors in the natural sciences, we will find more of the shared perspective between the natural sciences and the human sciences. In other words, the humanities should make use of methods from the natural sciences, while the natural sciences must learn from the human sciences. In doing so, it is possible that natural and human sciences will be unified into one "human knowledge."

The demonstration above shows not only the methodological unity between the natural sciences and the human sciences, but also the unification of scientism and idealism. While it reveals the unification of science and humanity as scientists have expected, it does acknowledge idealists' position that interpretation and explanation are different in form; however, it has shown that their differences in form are inadequate to justify the separation of the natural sciences from the humanities.

Notes

1 This chapter makes a distinction between explanation and interpretation in advance, and then discusses whether they are the same thing. I here want to discuss the methodological unity of the natural sciences, the social sciences, and the humanities (the latter two disciplines are alternately called "human sciences" according to Charles Taylor), but for the convenience of writing, this chapter will put social sciences and the humanities together in the discussion, ignoring the

distinctions between them. In contrast with natural sciences, which study natural phenomena, human sciences (social sciences and the humanities) study human behaviors.

2 Dray raises his own criticism to Hempel's revision, and the author's response to Dray can be found in Section 3 "Covering law thesis and history?" of this chapter.

3 Professor Yuankang Shih was my teacher at the Department of Philosophy, the Chinese University of Hong Kong. His academic research and personality greatly influenced me. My PhD dissertation *Relativism* aimed to tackle the issue of relativism with which he has long been concerned. I would like to express my deep gratitude to him here.

4 Salmon regards this kind of explanation as causal explanation as well, but his effort actually expands the content of causal explanation, which is no longer standard usage (Salmon, 1998, pp. 241–263).

5 Someone may worry that some bad policies such as exhausting all the resources to develop military power may sometimes be popular in some particular situation. For example, Adolf Hitler's military policy at the beginning of the Third Reich of Germany won huge support from the people. I would like to use chemical laws as an analogy to respond to these comments: some chemicals are poisonous, but if combined with other chemicals, certain combined products may be harmless to human beings.

6 Yuankang Shih also thinks the justification of natural sciences has a set of objective procedures, so they are objective. In fact, the scientific methods in the context of justification, from Carnap's inductive logic and Popper's hypothesis-falsification method to Bayesianism, have not succeeded so far. Nowadays, most philosophers of science have given up this kind of scientific methodology. He also mentions two ideas – "the choice of topic" and "the construction of object" – to argue that the social sciences are not as objective as the natural sciences. I have two responses. First, the choice of topic in the natural sciences is selective, too. Second, objects of the natural sciences also rely on value judgments of the scientific community, not just intuitive forms and intellectual categories. In this sense, objects of the natural sciences are constructed. Social constructivism even regards scientific knowledge as the result of construction by society, culture, and history. Moreover, laws in the social sciences can be as real as the laws in the natural sciences. For example, the disasters resulting from "the Great Leap Forward" movement that violated economic laws are as objective as the power of atomic bombs whose production is based on laws in the natural sciences.

9   Synthesis

Explanation, laws, and causation

All three concepts, scientific explanation, laws of nature, and causation, are central issues in the philosophy of science. At the Philosophy of Science Association's (PSA) 2008 Biennial Meeting, the problems of induction and causation–law–explanation were the two most important topics within the general philosophy of science.

Explanation, laws, and causation are closely connected, too. For instance, Hempel's scientific explanation models must include scientific laws and are thus called covering law models. We usually think only laws of nature can explain phenomena, while accidental generalizations have no such explanatory role. Between laws and causation, we must ask, which comes first? Is there singular causation without regularity? Should nomic expectability or causal mechanism be the most important in scientific explanation? These problems are all intertwined and interrelated, forming a complex problem cluster.

The author has, within the pages of this work, studied the relations between explanation, laws, and causation, proposed his points and arguments, and finally achieved a unified and comprehensive view. In addition, the author has examined several related issues, for example, the connection between explanation and reduction, the relation between explanation and interpretation, etc. In doing so, this research can hopefully play a positive role in resolving the central problem cluster of explanation–law–causation in the general philosophy of science.

To summarize, in the first chapter, "Hempel's scientific explanation models and their problems," the author introduces Hempel's three kinds of scientific explanation models: the Deductive-Nomological (DN) model, the Inductive-Statistical (IS) model, and the Deductive-Statistical model, as well as their variations: elliptical explanation, partial explanation, and explanation sketch. However, as the chapter describes, Hempel's explanation models meet a series of challenges, such as the asymmetry problem, the irrelevance problem, the requirement of high probability, and the requirement of maximal specificity.

In the second chapter, "Six decades of scientific explanation," the author reviews Salmon's three conceptions in the development of scientific explanation after Hempel. (1) The epistemic conception can upholds that "explaining an event is equal to showing its nomic expectability," following (revising and consummating) the epistemic approach pioneered by C. G. Hempel. Examples of this conception are van Fraassen's pragmatics of explanation and Philip Kitcher's unificationist models. (2) The modal conception holds that "scientific explanations aim at showing that the explained events must happen." For example, D. H. Mellor and others have proposed a modal interpretation for probabilistic explanations. (3) Finally, the ontic conception maintains that "scientific explanations aim at revealing causality and the intrinsic mechanism of the explained phenomena, and clarifying the status of it in the overall natural picture and hierarchical structure." Two examples of the ontic conception are W. Salmon's causal theory and P. Railton's DNP model.

The author suggests that to study scientific explanation, the crucial step is to understand scientific laws. Hempel's explanation models are also called covering law models, and their essential feature is that all scientific explanations must cover at least one scientific law. But what is the nature of laws? Hempel tries to provide a general form of lawlike sentences, but he finally fails. Thus, scientific law, the central concept in scientific explanation, is itself ambiguous, whereas the establishment of explanation models ultimately depends upon our understanding of scientific laws.

In Chapter 3, "The very nature of laws of nature," the author analyzes the two main approaches to the discussion of laws of nature, the regularity approach and the necessitarian approach. The regularity approach regards laws of nature as the real descriptions of how objects behave in actuality, while the necessitarian approach assumes that laws of nature not only describe what the world is, but also claim what the world must be. With that said, while the regularity approach meets the problem of opacity, the problem of confirmation, and confusion between epistemology and ontology, the necessitarian approach can never provide a satisfactory definition of necessity and thus falls prey to the identification problem, the inference problem, and non-instance laws. Van Fraassen and Giere both believe there are no laws of nature. However, they simply change the problem of laws into either symmetry or equations (or principles), and, in so doing, they never solve or dissolve the problem. Mitchell's approach is more plausible, but her denial of the dichotomy between genuine laws of nature and accidental generalizations is not convincing.

The author suggests that the necessity of scientific laws is stipulated by nature, which is an ontological problem, and the necessitarian approach provides a good solution; at the same time, while laws of nature form the best and most coherent system we can use to explain and promote the world, the regularity approach provides a good option for the epistemological approach. Therefore, we can come back to Hume and combine the regularity approach and the necessitarian approach without conflict or contradiction.

In Chapter 4, "The conceptions of scientific explanation and approaches to laws of nature," the author reviews main conceptions of scientific explanation and two approaches to laws of nature, arguing: (1) van Fraassen's pragmatics of explanation is just an important supplement to Hempel's logic of explanation. If we further ask what his concept of relevant relation is, we may finally end up appealing to laws. Kitcher's unificationist account perfectly matches with the MRL view in the regularity approach to laws. (2) Mellor holds a modal interpretation of probabilistic explanation, but that is inconsistent with his regularity approach to laws. (3) Salmon insists that causal or lawful nomicity is essential in scientific explanation, and his ontic conception should be compatible with Hempel's covering law thesis. Railton's DNP model requires laws, and his conception of laws belongs to the regularity approach. (4) Watkins' deductivism is even more rigidly insistent on general laws or probability laws. (5) Cartwright thinks all laws are *ceteris paribus*, so she proposes the informal simulacrum account of explanation.

At the end of the fourth chapter, the author draws two conclusions. (1) The conceptions of scientific explanation are closely related to approaches to laws of nature. Comparatively speaking, the unificationist model of explanation maximally fits with the regularity approach to laws of nature. (2) Laws are essential in scientific explanation, so Hempel's covering law thesis is still an important "dogma" (doctrine) of empiricism.

In Chapter 5, "Causal mechanism and lawful explanation," the author tackles Salmon's causal mechanism account of scientific explanation. Salmon thinks causation is prior to explanation, and there is no explanation without causation. In addition, he argues that some causation can be non-lawful, so causation is more basic than laws. In light of this, the author provides four arguments against him: (1) Salmon uses "mark transmission" as the condition distinguishing the causal process from the pseudo process, but the concept itself is causal, so he encounters the problem of circular definition. (2) Salmon accepts singular causation, but the author thinks we can identify those singular causations only if we have some background of scientific laws. (3) Causation can be expressed in the form of laws, yet some laws in physics cannot be expressed in the form of causation – for example, the correlation of identical particles in quantum mechanics can be described with a law, but there is no action-at-distance causation between the two particles. (4) Salmon thinks scientific explanation must be ontic, but as a kind of human activity, explanation should be epistemic. Ontologically, physical behaviors can follow Salmon's bottom-up model, but epistemologically, scientific explanation should use the top-down model. In addition, the author proposes, the ontological structure of the world should be described from events (or processes) to causation to regularity (laws) and, finally, to science system, but the epistemological order should be from science system to laws to explanation and causation.

In Chapter 6, "Is there such a thing as a *ceteris paribus* law?" the author tries to argue for *ceteris paribus* (CP) laws. John Earman, J. Roberts, and S. Smith provide several arguments against CP laws: (1) CP clauses can be easily eliminated if we use scientific language properly. (2) We cannot substitute testable auxiliaries for the CP clauses, so there are no ways to test CP laws. (3) The so-called CP laws are just differential equations of evolution type, but laws are strict, that is, universal and necessary. However, the author raises the following objections to ERS: (1) CP clauses may involve infinite kinds of interferences; therefore, they can be ineliminable even when using scientific terminology. (2) It is possible to test the contraposition of a CP law and, thus, the law itself. (3) Earman's account of differential equations may be inconsistent with his MRL view of laws of nature. Differential equations can be laws. The author's thought experiment shows that Earman's distinction between conditions for the truth of a law and conditions for the validity of its application may be inconsistent with his supervenience thesis. In summary, the author recommends giving up residual Platonic idealism in the modern sciences and is cautiously optimistic about CP laws.

In Chapter 7, "Explanation and reduction," the author discusses the relation between explanation and reduction. Explanation and reduction have a similar logical structure, so E. Nagel uses Hempel's explanation models to represent theory reduction. The author clarifies a series of concepts such as language reduction, theory reduction, discipline reduction, micro-reduction, ontological reduction, methodological reduction, and explanatory reduction.

Then, the author remarks upon the related types of reductionism. (1) Language reductionism, in attempting to unify all discipline languages, is of practical significance but meets the holism problem. (2) Theory reduction, discipline reduction, and micro-reduction are closely related, and their requirements are higher than generally argued. Since theory reductionism is problematic, discipline reductionism and micro-reductionism, which are both based on theory reductionism, are seen to be improbable. (3) Ontological reductionism, which is also called "compositional materialism," is the most acceptable reductionism in academic circles. (4) Methodological reductionism, also called "decomposition strategy," is replete with practical significance. However, the author reminds readers not to ignore mutual learning between disciplines. (5) Epistemic reduction includes theory reduction and explanatory reduction. Since theory reductionism and explanatory reductionism are both fraught with many difficulties, epistemic reductionism should also be considered infeasible.

In Chapter 8, "Scientific explanation and historical interpretation," the author examines the relation between explanation and interpretation. Hempel wishes to apply his explanation models to the human sciences, thereby achieving the methodological unity of empirical sciences.; however, his effort meets criticism from many philosophers and historians. The author tries to defend Hempel with the following arguments. (1) The deductive thesis is not fundamental to scientific explanation, while the covering law thesis is essential. (2) There are laws in historical studies, and the covering law thesis is necessary for historical explanations. (3) The natural sciences and the human sciences both require value judgments of academic communities and both do not "see" the world, but rather "see as." (4) Finally, all empirical sciences involve both interpretation and explanation; hence, the distinction between interpretation and explanation cannot justify the separation between the natural sciences and the human sciences.

The author argues that the natural sciences and the human sciences can be methodologically unified. That being said, this kind of unity is not achieved by the natural sciences unifying the human sciences; instead, with the realization of more human factors in the natural sciences, we will find more shared perspectives between the natural sciences and the human sciences. Thus, through mutual learning, the natural and human sciences will be unified into one "human knowledge."

This book aims to achieve some kind of holistic synthesis by arguing that our scientific knowledge is a holistic system and important component of which is law. With scientific laws, we can explain and predict phenomena in the world, as well as identify causations in both science and ordinary life. All three concepts – explanation, law, and causation – can be understood broadly, thus the three concepts can be applied in both the natural sciences and the human sciences. Hopefully, the boundary between nature and humanity can be opened up, and we can finally achieve a unified system of science.

Bibliography

Achinstein, P. (1983). *The nature of explanation*. New York: Oxford University Press.

Ayer, A. J. (1936). *Language, truth, and logic*. London: Penguin.

Ayer, A. J. (1956). What is a law of nature? *Revue Internationale de Philosophie, 10*, 144–165.

Balashov, Y., and Rosenberg, A. (Eds.) (2002). *Philosophy of science: Contemporary readings*. London and New York: Routledge.

Bernstein, R. J. (1988). *Beyond objectivism and relativism*. Philadelphia, PA: University of Pennsylvania Press.

Bloor, D. (1991). *Knowledge and social imagery*. Chicago, IL: The University of Chicago Press.

Boyd, R. et al. (Eds.) (1991). *The philosophy of science*. Cambridge, MA: MIT Press.

Braybrooke, D. (1987). *Philosophy of social science*. Newark, NJ: Prentice-Hall.

Brigandt, I., and Love, A. (2008). Reductionism in biology. In *Stanford encyclopedia of philosophy*. Retrieved from http://plato.stanford.edu/entries/reduction-biology

Bromberger, S. (1966). Why-questions. In R. G. Colodny (Ed.), *Mind and cosmos*, pp. 86–108. Pittsburgh, PA: University of Pittsburgh Press.

Brown, H. I. (1990). Prospective realism. *Studies in History and Philosophy of Science, 21*, 211–42.

Browne, M. E. (1999). *Schaum's outline of theory and problems of physics for engineering and science* (Series: Schaum's Outline Series). London: McGraw-Hill Companies.

Burtt, E. A. (1932). *The metaphysical foundations of modern physical science*. New York: Humanity Books.

Cao, T. Y. 曹天予. 科学和哲学中的后现代性. 曹南燕译. 哲学研究, 2000(2).

Carnap, R. (1938). Logical foundation of the unity of science. In R. Boyd et al. (Eds.) (1991). *The philosophy of science*, pp. 393–404. Cambridge, MA: MIT Press.

Carnap, R. (1950). *Logical foundation of probability*. Chicago, IL: The University of Chicago Press.

Carnap, R. (1995). *An introduction to the philosophy of science*. M. Gardner (Ed.). New York: Dover.

Carroll, J. (Ed.) (2004). *Readings on laws of nature*. Pittsburgh, PA: University of Pittsburgh Press.

Cartwright, N. (1983). *How the laws of physics lie*. Oxford: Clarendon.

Cartwright, N. (1989). *Nature's capacities and their measurement*. Oxford: Oxford University Press.

Cartwright, N. (1993). In defence of 'this worldly' causality: Comments on van Fraassen's laws and symmetry. *Philosophy and Phenomenological Research, 53*(2): 423–29.

Cartwright, N. (1998). Where do laws of nature come from? In P. Matthias (Hrsg.), *Nancy Cartwright: Laws, capacities and science: Vortrag und Kolloquium in Münster 1998*, pp. 1–30. Münster: LIT-Verlag.

Cartwright, N. (1999). *The dappled world*. Cambridge: Cambridge University Press.

Cartwright, N. (2002). In favor of laws that are not *Ceteris Paribus* after all. *Erkenntnis, 57*, 425–439.

Chen, B. 陈波. 休谟问题和金岳霖的回答. 中国社会科学, 2001(3).

Chen, X. 陈晓平. 主观主义概率论对于休谟问题的"解决". 自然辩证法通讯, 1994(1).

Chen, X. 陈晓平. 归纳逻辑与归纳悖论. 武汉: 武汉大学出版社, 1994.

Chen, X. 陈晓平. 贝叶斯认证逻辑及其应用. 自然辩证法研究, 1994(9).

Chen, X. 陈晓平. 大弃赌定理及其哲学意蕴. 自然辩证法通讯, 1997(2).

Chen, X. 陈晓平. 关于归纳逻辑的若干问题. 自然辩证法通讯, 2000(5).

Churchland, P. M., and Hooker, C. A. (Eds.) (1985). *Images of science*. Chicago, IL: The University of Chicago Press.

Cohen, J. (1989). *An introduction to the philosophy of induction and probability*. Oxford: Clarendon Press.

Collingwood, R. G. (1948). *The idea of history*. Oxford: Clarendon Press.

Colodny, R. G. (Ed.) (1966). *Mind and cosmos*. Pittsburgh, PA: University of Pittsburgh Press.

Corfield, D., and Williamson, J. (2001). *Foundations of Bayesianism*. Dordrecht: Kluwer Academic Publishers.

Couvalis, G. (1997). *The philosophy of science: Science and objectivity*. London: Sage.

Curd, M., and Cover, J. A. (Eds.) (1998). *Philosophy of science: The central issues*. New York: W. W. Norton & Co.

Dainian, F. 范岱年. 一部为科学实在论作辩护的当代物理学思想史. 自然辩证法研究, 1998(1).

Donagan, A. (1959). Explanation in history. In P. Gardiner (Ed.), *Theory of history*, pp. 428–43. New York: The Free Press.

Dray, W. H. (1957). *Laws and explanation in history*. London: Oxford University Press.

Dray, W. H. (1964). *Philosophy of history*. Upper Saddle River, NJ: Prentice-Hall.

Dretske, F. I. (1977). Laws of nature. *Philosophy of Science, 44*, 248–268.

Dubuc, J. (Ed.) (1993). *Philosophy of probability*. Dordrecht: Kluwer Academic Publishers.

Earman, J. (1992). *Bayes or bust? A critical examination of Bayesian confirmation theory*. Cambridge, MA: MIT Press.

Earman, J. (2002). Laws of nature. In Y. Balashov and A. Rosenberg (Eds.), *Philosophy of science: Contemporary readings*, pp. 116–17. Washington, DC: Psychology Press.

Earman, J. and Roberts, J. (1999). *Ceteris Paribus*: There is no problem of provisos. *Synthese, 118*, 439–478.

Earman, J., Roberts, J., and Smith, S. (2002). *Ceteris Paribus* lost. *Erkenntnis, 57*, 281–303.

Earman J., et al. (Eds.) (2002). *Ceteris Paribus laws*. Norwell, MA and Dordrecht: Kluwer Academic Publishers.

Einstein, A. (1961). *Relativity*. R. W. Lawson (Trans.). New York: Three Rivers Press.

Elgin, M., and Sober, M. (2002). Cartwright on explanation and idealization. *Erkenntnis, 57*, 441–450.

Feigl, H., and Maxwell, G. (Eds.) (1962). *Scientific explanation, space, and time*. Minneapolis, MN: University of Minnesota Press.

Fetzer J. H. (Ed.) (2000). *Science, explanation, and rationality: The philosophy of Carl G. Hempel*. New York: Oxford University Press.

Fetzer J. H. (2010). Carl Hempel. In *Stanford encyclopedia of philosophy*. Retrieved from http://plato.stanford.edu/entries/hempel/

Feyerabend, P. (1963). How to be a good empiricist: A plea for tolerance in matters epistemological. In M. Curd and J. A. Cover (Eds.) (1998). *Philosophy of science: the central issues*, pp. 922–49. New York and London: W. W. Norton & Co.

Feyerabend, P. (1993). *Against method* (3rd ed.). London: Verso.

Friedman, M. (1971). Explanation and scientific understanding. *The Journal of Philosophy, 71*, 5–19.

Garfinkel, A. (1981). Reductionism. In R. Boyd et al. (Eds.) (1991). *The philosophy of science*, pp. 443–62. Cambridge, MA: MIT Press.

Gibson, Q. (1960). *The logic of social enquiry*. London: Routledge & K. Paul.

Giere, R. (1999). *Science without laws*. Chicago, IL: The University of Chicago Press.

Glymour, C. (1980). *Theory and evidence*. Princeton, NJ: Princeton University Press.

Glymour, C. (2002). A semantics and methodology for *Ceteris Paribus* hypothesis. *Erkenntnis, 57*, 395–405.

Goodman, N. (1955). *Fact, fiction, and forecast* (4th ed.). Cambridge, MA: Harvard University Press.

Hacking, I. (2001). *An introduction to probability and inductive logic*. Cambridge: Cambridge University Press.

Hahn, H. et al. (1929). The scientific conception of the world: The Vienna Circle. In S. Sarkar (Ed.) (1996). *The emergence of logical empiricism: From 1900 to the Vienna Circle*, pp. 321–41. New York: Garland Publishing.

Hempel, C. G. (1965). *Aspects of scientific explanation and other issues in the philosophy of science*. New York: The Free Press.

Hempel, C. G. (1966). *Philosophy of natural science*. Englewood Cliffs, NJ: Prentice-Hall.

Hempel, C. G. (1968). Explanation in science and in history. In P. H. Nidditch (Ed.), *The philosophy of science*, pp. 54–79. London: Oxford University Press.

Hempel, C. G. (1988). Provisos: A philosophical problem concerning the inferential function of scientific laws. In A. Grünbaum and W. Salmon (Eds.), *The limits of deductivism*, pp. 19–36. Berkeley, CA: University of California Press.

Hempel, C. G. (1998). Science and human values. In E. D. Klemke et al. (Eds.), *Introductory readings in the philosophy of science* (3rd ed.), pp. 499–514. New York: Prometheus.

Ho, H.-h. 何秀煌. 记号、意识与典范. 台北：东大图书公司，1999.

Horwich, P. (Ed.) (1993). *World change*. Cambridge, MA: MIT Press.

Howson, C., and Urbach, P. M. (1993). *Scientific reasoning: The Bayesian approach* (2nd ed.). Chicago, IL: Open Court.

Hume, D. (1985). *A treatise of human nature*. New York: Penguin Books.

Hume, D. (1993). *An enquiry concerning human understanding* (2nd ed.). New York: Hackett Publishing Company, Inc.

Hung, T. 洪谦. 论逻辑经验主义. 北京：商务印书馆，1999.

Hung, T. 洪谦. 逻辑经验主义论文集. 三联书店(香港)有限公司，1990.

Hung, T. 洪谦. 维也纳学派哲学. 北京：商务印书馆，1989.

Hung, T. 洪谦编. 逻辑经验主义(上下卷). 北京：商务印书馆，1982.

Jiang, T. 江天骥. 科学哲学名著选读. 武汉：湖北人民出版社，1988.

Kant, I. (1929). *Critique of pure reason*. N. K. Smith (Trans.). London: Macmillan.

Kitcher, P. (1981). Explanatory unification. *Philosophy of Science*, *48*, 507–531.

Kitcher, K. (1984). 1953 and all that: A tale of two sciences. In M. Curd and J. A. Cover (Eds.) (1998). *Philosophy of science: the central issues*, pp. 971–1003. New York and London: W. W. Norton & Company, Inc.

Kitcher, P. (1989). Explanatory unification and the causal structure of the world. In P. Kitcher and W. Salmon (Eds.), *Scientific explanation*, pp. 410–505. Minneapolis, MN: University of Minnesota Press.

Klee, R. (1999). *Scientific inquiry: Readings in philosophy of science*. Oxford: Oxford University Press.

Klemke, E. D. et al. (Eds.) (1998). *Introductory readings in the philosophy of science* (3rd ed.). New York: Prometheus Books.

Koertge, N. (2000). "New Age" philosophies of science: Constructivism, feminism and postmodernism. *British Journal for Philosophy of Science*, *51*, 667–83.

Koyre, A. (1978). *Galileo studies*. Atlantic Highland, NJ: Humanities Press.

Kraut, R. (2013). Plato. In *Stanford encyclopedia of philosophy*. Retrieved from http://plato.stanford.edu/entries/plato/

Kuhn, T. (1957). *The Copernican revolution*. Cambridge, MA: Harvard University Press.

Kuhn, T. (1970). *The structure of scientific revolutions* (2nd ed.). Chicago, IL: University of Chicago Press.

Kuhn, T. (1977). *The essential tension*. Chicago, IL: The University of Chicago Press.

Kuhn, T. (1982). Commensurability, comparability, and communicability. *PSA*, *1982*(2): 669–88.

Kuhn, T. (1998). The natural and the human sciences. In E. D. Klemke et al. (Eds.), *Introductory readings in the philosophy of science* (3rd ed.). New York: Prometheus.

Kuhn, T. (2000). *The road since structure*. J. Conant and J. Haugeland (Eds.). Chicago, IL: The University of Chicago Press.

Lakatos, I. (1978). *The methodology of scientific research programmes: Philosophical papers. Vol. I*. J. Worrall and G. Currie (Eds.). Cambridge: Cambridge University Press.

Lakatos, I., and Musgrave, A. (Eds.) (1970). *Criticism and the growth of knowledge*. Cambridge: Cambridge University Press.

Lange, M. (1993). Natural laws and the problem of provisos. *Erkenntnis*, *38*, 233–48.

Lange, M. (2002). Who's afraid of *Ceteris-Paribus* law? Or: How I learned to stop worrying and love them. *Erkenntnis*, *57*, 407–423.

Laudan, L. (1977). *Progress and its problems*. London: Routledge & K. Paul.

Laudan, L. (1984). *Science and value*. Berkeley, CA: University of California Press.

Laudan, L. (1990). *Science and relativism*. Chicago, IL: The University of Chicago Press.

Laudan, L. (1996). *Beyond positivism and relativism*. Boulder, CO: Westview Press.

Leplin, J. (Ed.) (1984). *Scientific realism*. Berkeley, CA: University of California Press.

Lewis, D. (1973). *Counterfactuals*. Cambridge, MA: Harvard University Press.

Lin, C.-h. 林正弘. 伽利略、波柏、科学说明. 台北：东大图书股份有限公司，1989

Lipton, P. (1999). All else being equal. *Philosophy, 74*, 155–168.

McCarthy, T. (1977). On an Aristotelian model of scientific explanation. *Philosophy of Science, 44*, 159–166.

McErlean, J. (2000). *Philosophies of science: From foundations to contemporary issues*. Belmont, CA: Wadsworth Publishing Co.

Mellor, D. H. (1976). Probable explanation. *Australasian Journal of Philosophy, 54*(3), 231–241.

Mellor, D. H. (1998). Necessities and universals in natural laws. In M. Curd and J. A. Cover (Eds.), *Philosophy of science: the central issues*, pp. 846–864. New York: W. W. Norton & Co.

Mill, J. S. (1904). *A system of logic*. New York: Harper and Row.

Mitchell, S. (1997). Pragmatic laws. *Philosophy of Science, 64*(4), 468–479.

Mitchell, S. (2000). Dimensions of scientific law. *Philosophy of Science, 67*(2), 242–265.

Mitchell, S. (2002). *Ceteris Paribus*: An inadequate representation for biological contingency. *Erkenntnis, 57*, 329–350.

Mitchell, S. (2008). The PPT of the course "Philosophy of Biology" at the first "Pitt-Tsinghua Summer School for Philosophy of Science" on August 18–22, 2008, co-organized by the Institute of Science, Technology and Society, Tsinghua University and the Center for Philosophy of Science, the University of Pittsburgh.

Morreau, M. (1999). Other things being equal. *Philosophical Studies, 96*, 163–182.

Morton, A. (2000). Saving epistemology from the epistemologists: Recent work in the theory of knowledge. *British Journal for Philosophy of Science, 51*, 685–704.

Nagel, E. (1961). *The structure of science: Problems in the logic of scientific explanation*. London: Routledge & Kegan Paul.

Nagel, E. (1974). Issues in the logic of reductive explanations. In M. Curd and J. A. Cover (Eds.) (1998). *Philosophy of science: The central issues*, pp. 905–21. New York and London: W. W. Norton & Co.

Nagel, E., and Brandt, R. B. (Ed.) (1965). *Meaning and knowledge*. Orlando, FL: Harcourt, Brace & World.

Newton-Smith, W. H. (1981). *The rationality of science*. London: Routledge.

Nidditch, P. H. (Ed.) (1968). *The philosophy of science*. London: Oxford University Press.

Nozick, R. (1993). *The nature of rationality*. Princeton, NJ: Princeton Universtiy Press.

Nye, A. (Ed.) (1998). *Philosophy of language: The big questions*. Boston, MA: Blackwell.

Oppenheim, P., and Putnam, H. (1958). Unity of science as a working hypothesis. In R. Boyd et al. (Eds.) (1991). *The philosophy of science*, pp. 905–21. Cambridge, MA: MIT Press.

Palmer, R. E. (1969). *Hermeneutics*. Evanston, IL: Northwestern University Press.

Persky, J. (1990). *Ceteris Paribus. Journal of Economic Perspectives*, *4*, 187.

Polanyi, M. (1964). *Personal knowledge*. Chicago, IL: The University of Chicago Press.

Popper, K. (1963). *Conjectures and refutations*. London: Routledge & Kegan Paul.

Popper, K. (1972). *Objective knowledge*. Oxford: Clarendon Press.

Pryor, J. (2001). Highlights of recent epistemology. *British Journal for Philosophy of Science*, *52*, 95–124.

Psillos, S. (2000). The present state of the scientific realism debate. *British Journal for the Philosophy of Science*, *51*(4): 705–728.

Psillos, S. (2002). *Causation and explanation*. Kingston, ON: McGill-Queen's University Press.

Putnam, H. (1975). *Mathematics, matter and method*. Cambridge: Cambridge University Press.

Putnam, H. (1981). *Reason, truth, and history*. Cambridge: Cambridge University Press.

Quine, W. V. O. (1960). *Word and object*. Cambridge, MA: MIT Press.

Quine, W. V. O. (1980). *From a logical point of view*. Cambridge, MA: Harvard University Press.

Railton, P. (1978). A deductive-nomological model of probabilistic explanation. *Philosophy of Science*, *45*, 206–226.

Railton, P. (1981). Probability, explanation, and information. *Synthese*, *48*, 233–256.

Ramsey, F. (1978). *Foundations of mathematics*. Atlantic Highlands, NJ: Humanities Press.

Reichenbach, H. (1947). *Elements of symbolic logic*. London: Macmillan.

Reichenbach, H. (1951). *The rise of scientific philosophy*. Berkeley, CA: University of California Press.

Rescher, N. (1970). *Scientific explanation*. New York: Free Press.

Rescher, N. (1988). *Rationality*. Oxford: Clarendon Press.

Reutlinger, A., Schurz, G., and Hüttemann, A. (2011). *Ceteris Paribus* laws. In *Stanford encyclopedia of philosophy*. Retrieved from https://plato.stanford.edu/entries/ceteris-paribus/

Rorty, R. (1991). *Objectivity, relativism, and truth*. Cambridge: Cambridge University Press.

Ruben, D. (1990). *Explaining explanation*. New York: Routledge.

Salmon, W. (1984). *Scientific explanation and the causal structure of the world*. Princeton, NJ: Princeton University Press.

Salmon, W. (1988). *Causality and explanation*. Oxford: Oxford University Press.

Salmon, W. (1989). *Four decades of scientific explanation*. Minneapolis, MN: University of Minnesota Press.

Salmon, W. (1998). Scientific explanation: How we got from there to here. In E. D. Klemke et al. (Eds.), *Introductory readings in the philosophy of science* (3rd ed.), pp. 241–65. New York: Prometheus.

Sankey, H. (1997). *Rationality, relativism and incommensurability*. London: Ashgate Publishing, Ltd.

Sarkar, S. (Ed.) (1996). *The emergence of logical empiricism: From 1900 to the Vienna Circle*. New York: Garland Publishing.

Scheffler, I. (1964). *The anatomy of inquiry*. London: Routledge & Kegan Paul, Ltd.

Schilpp, P. A. (Ed.) (1963). *The philosophy of Rudolf Carnap*. Chicago, IL: Open Court.

Schilpp, P. A. (1974). *The philosophy of Karl Popper*. Chicago, IL: Open Court.

Schrödinger, E. (1944). *What is life?* Cambridge: Cambridge University Press.

Schurz, G. (2002). *Ceteris Paribus* laws: Classification and deconstruction. *Erkenntnis, 57*, 351–372.

Scriven, M. (1959). Definitions, explanations, and theories. In H. Feigl, M. Scirven, and G. Maxwell (Eds.), *Minnesota studies in the philosophy of science* (Vol. 2), pp. 99–195. Minneapolis, MN: University of Minnesota Press.

Scriven, M. (1962). Explanations, predictions, and laws. In H. Feigl and G. Maxwell (Eds.), *Minnesota studies in the philosophy of science* (Vol. 3), pp. 170–230. Minneapolis, MN: University of Minnesota Press.

Scriven, M. (1963). The temporal asymmetry between explanations and predictions. In B. Baumrin (Ed.), *Philosophy of science: The Delaware seminar* (Vol. 1), pp. 97–106. New York: John Wiley.

Searle, J. (1995). *The construction of social reality*. New York: Free Press.

Searle, J. (1999). *Mind, language and society*. London: Phoenix.

Sellars, W. (1962). *Science, perception and reality*. New York: Humanities Press.

Shapere, D. (1984). *Reason and the search for knowledge*. Dordrecht: Reidel.

Shi, Y. 施雁飞. 科学解释学. 长沙：湖南出版社，1991.

Shih, Y. 石元康. 意义与社会科学的客观性. 食货月刊复刊第12卷第7期，1982年10月.

Shih, Y. 石元康. 实证论与历史说明. 史学评论. 第6期，1983年9月.

Shih, Y. 石元康. 历史中的原因、目的与理由". 鹅湖. (第100期，1983年10月.

Shu, W. and Qiu, R. 舒炜光、邱仁宗主编. 当代西方科学哲学述评. 北京：人民出版社，1987.

Smart, J.J.C. (1968). *Between science and philosophy*. New York: Random House.

Sphohn, W. (2002). Laws, *Ceteris Paribus* conditions, and the dynamics of belief. *Erkenntnis, 57*, 373–394.

Stove, D. (1982). *Popper and after: Four modern irrationalists*. Oxford: Pergamon Press.

Suppe, F. (1977). *The structure of scientific theories*. Champaign, IL: University of Illinois Press.

Table of contents. In *Stanford encyclopedia of philosophy*. Retrieved from http://plato.stanford.edu/contents.html#r

Taylor, C. (1998). Interpretation and the sciences of man. In E. D. Klemke et al. (Eds.), *Introductory readings in the philosophy of science* (3rd ed.), pp. 110–27. New York: Prometheus.

Thagard, P. (1988). *Computational philosophy of science*. Cambridge, MA: MIT Press.

Thomas, D. (1979). *Naturalism and social science.* Cambridge: Cambridge University Press.

van Fraassen, B. (1977). The pragmatics of scientific explanation. *American Philosophical Quarterly, 1977, 14*(2), 143–150.

van Fraassen, B. (1980). *The scientific image.* New York: Oxford University Press.

van Fraassen, B. (1989). *Laws and symmetry.* Oxford: Oxford University Press.

Wang, W. 王巍. 相对主义. 北京: 清华大学出版社, 2003.

Wang, W. 王巍. 科学哲学问题研究. 北京: 清华大学出版社, 2004.

Watkins, J. (1984). *Science and skepticism.* Princeton, NJ: Princeton University Press.

Weatherford, R. (1982). *Philosophical foundations of probability theory.* London: Routledge & Kegan Paul.

Winch, P. (1990). *The idea of social science and its relation to philosophy* (2nd ed.). London: Routledge.

Windelband, W. (1980). History and natural science. G. Oakes (Trans.) *History and Theory 19*, 169–85.

Wittgenstein, L. (1961). *Tractatus logico-philosophicus.* D.F. Pears and B.F. McGuinness (Trans.). London: Routledge.

Wittgenstein, L. (1967). *Philosophical investigations.* G.E.M. Anscombe (Trans.). Oxford: Basil Blackwell.

Woodward, J. (1989). The causal mechanical model of explanation. In *Minnesota studies in the philosophy of science*, pp. 357–83. Minneapolis, MN: University of Minnesota Press.

Woodward, J. (2002). There is no such thing as a *Ceteris Paribus* law. *Erkenntnis, 57*, 303–328.

Woodward, J. (2009). Scientific explanation. In *Stanford encyclopedia of philosophy*. Retrieved from https://plato.stanford.edu/entries/scientific-explanation/

Ye, C. 叶闯. 亨普尔科学解释模型的核心问题及其解决. 自然辩证法通讯, 1997(4).

Zheng, X. 郑祥福. 西方科学哲学中的"科学说明"的研究走向. 哲学动态, 1997(3).

Zheng, X. 郑祥福. 论范·弗拉森语用学的科学说明观. 自然辩证法通讯, 1998(4).

Zhuang, H. 张华夏. 科学解释标准模型的建立、困难与出路. 科学技术与辩证法, 2002(1).

Index